T0396973

"The tremendous value of the approach laid out in *Experiential Learning and Community Partnerships for Sustainable Development* is that it gives students the opportunity to experience their own power and agency to impact the world in a positive way."

Sally Crimmins Villela, *Associate Vice Chancellor for Global Affairs and Senior International Officer, The State University of New York*

"*Experiential Learning and Community Partnerships for Sustainable Development* is a treasure. It's like a handbook for lost students like me. If only I had this in my freshman year."

Muhammed Moinuddin, *Mechanical Engineering student, University at Buffalo*

Experiential Learning and Community Partnerships for Sustainable Development

This book addresses the growing demand for applied experiences that move students beyond learning into the realm of doing by supporting the development of skills and competencies that align with emerging areas of innovation and work. It considers the urgent need to promote and invest in skills that support sustainable development, such as those needed to analyze and mitigate climate change. The authors argue that this challenge provides an opportunity to reimagine the use of Experiential Learning, connecting students with community-based partners doing the work of sustainable development around the world. Featuring compelling case studies of project partners in Nigeria, Uganda, and Tanzania working to address the complexities of climate change, they offer a practical model for implementing Experiential Learning that can be translated and scaled across sectors and resource environments. It is aimed at scholars and educators working across higher education and international education with interests in digital and experiential education.

Mara Huber is Senior Director of Instructional Innovation and Transformation at the University at Buffalo Office of Curriculum, Assessment and Teaching Transformation (CATT).

Michael Jabot is SUNY Distinguished Professor of Science Education at the State University of New York at Fredonia, USA.

Christina Heath is Administrative Director of the Experiential Learning Network, University at Buffalo, USA.

Routledge Research in Higher Education

Identity Construction as a Spatiotemporal Phenomenon within Doctoral Students' Intellectual and Academic Identities
Contradictions, Contestations and Convergences
Rudo F. Hwami

An International Approach to Developing Early Career Researchers
A Pipeline to Robust Education Research
Edited by Stephen Gorard and Nadia Siddiqui

The Development of University Teaching Over Time
Pedagogical Approaches from 1800 to the Present
Tom O'Donoghue

The Layered Landscape of Higher Education
Capturing Curriculum, Diversity and Cultures of Learning in Australia
Edited by Margaret Kumar, Supriya Pattanayak and Nish Belford

Authority, Passion, and Subjected-Centered Teaching
A Christian Pedagogical Philosophy
Christopher J. Richmann

Experiential Learning and Community Partnerships for Sustainable Development
A Foundational Model for Climate Action
Mara Huber, Michael Jabot, and Christina Heath

For more information about this series, please visit: www.routledge.com/Routledge-Research-in-Higher-Education/book-series/RRHE

Experiential Learning and Community Partnerships for Sustainable Development

A Foundational Model for Climate Action

Mara Huber, Michael Jabot, and Christina Heath

NEW YORK AND LONDON

First published 2025
by Routledge
605 Third Avenue, New York, NY 10158

and by Routledge
4 Park Square, Milton Park, Abingdon, Oxon, OX14 4RN

Routledge is an imprint of the Taylor & Francis Group, an informa business

Library of Congress Cataloging-in-Publication Data
Names: Huber, Mara, author. | Jabot, Michael., author. | Heath, Christina, author.
Title: Experiential learning and community partnerships for sustainable development : a foundational model for climate action / Mara Huber, Michael Jabot and Christina Heath.
Description: New York, NY : Routledge, 2025. |
Series: Routledge research in higher education |
Includes bibliographical references and index.
Identifiers: LCCN 2024026255 (print) | LCCN 2024026256 (ebook) |
ISBN 9781032741710 (hardback) | ISBN 9781032788050 (paperback) |
ISBN 9781003489337 (ebook)
Subjects: LCSH: Experiential learning. | Community and college. |
Climatic changes–Study and teaching (Higher) |
Sustainable development–Study and teaching (Higher)
Classification: LCC LB1027.23 .H83 2025 (print) |
LCC LB1027.23 (ebook) | DDC 370.15/23–dc23/eng/20240705
LC record available at https://lccn.loc.gov/2024026255
LC ebook record available at https://lccn.loc.gov/2024026256

ISBN: 978-1-032-74171-0 (hbk)
ISBN: 978-1-032-78805-0 (pbk)
ISBN: 978-1-003-48933-7 (ebk)

DOI: 10.4324/9781003489337

Typeset in Times New Roman
by Newgen Publishing UK

Contents

Preface

When it comes to Experiential Learning, intentions matter. When students approach a Project Challenge with plans to develop specific skills or competencies, they are more likely to experience growth in those identified areas. Or put another way; they often get what they expect.

In sharing this book, we have specific intentions in mind. We hope to make a case for scaling Experiential Learning through Project Challenges, connecting students with community partners for collaborative projects focused on the Sustainable Development Goals (SDGs). But beyond sharing our approach, we seek to inspire readers to action, beginning to translate our model within their own contexts and environments, making Project Challenges accessible to their students while engaging local partners who are doing the work of sustainable development within vulnerable communities and ecosystems.

The structure of the book aligns closely with these intentions. The first section makes a case for Project Challenges through both pedagogical and practical lenses, and the second shares a scalable model for creating and supporting related engagement. Concrete examples are provided following each chapter in Part I along with guiding questions and activities to support design and translation work in Part II. The second half of the book is intended to move readers to action. In Part III we share case studies of partners and students who are engaging in SDG-related Project Challenges in Uganda, Tanzania, and Nigeria. The final chapter explores broader impacts and opportunities to connect Project Challenges with emerging areas of innovation including STEM Education, Geospatial Technology and Open Science, NASA GLOBE, and Restorative Agriculture.

While the book focuses specifically on SDG-related Project Challenges, it utilizes the Experiential Learning Network (ELN) model as its foundational approach. The ELN model was developed by authors Huber and Heath at the University at Buffalo (UB) in 2019 as a web-based system for connecting students with mentored projects and supporting engagement through PEARL modules and digital badges (Huber, Heath, Baxter, and Reed, 2021). SDG Project Challenges, a signature category of ELN projects, were first introduced in 2020 during the pandemic, when in-person Experiential Learning shut down and students were seeking opportunities for virtual engagement. As we sought to meet growing demand and looked to expand our portfolio of community-based partners, we drew on an existing mentoring relationship with YALI, the Young African Leaders Initiative, and an annual study abroad course to Tanzania that featured engagement with community-based partners who were leading humanitarian and environmental work. With the help of our partners, our network of community-based organizations quickly grew along with interest in translating the ELN model within diverse African contexts, as detailed in Part III.

In 2023 we received funding from our State University of New York (SUNY) system through an Innovative Instruction Technology Grant (IITG) to scale our SDG Project Challenges and digital badges, allowing engagement with non-UB students and faculty and their community-based partners. Although modest in scale, the pilot has demonstrated the practicality of connecting students, partners, and mentors – wherever they are – through collaborative Project Challenges focused on sustainable development. The implications are compelling. As we work to make Experiential Learning accessible to all students everywhere, we can transform our most pressing challenges into opportunities for both individual and collective impacts, while supporting and elevating the work of community-based partners. This is the promise of Experiential Learning and the intention for this book.

Reference

Huber, M., Heath, C., Baxter, C, & Reed, A. (2021). Using digital badges to design a comprehensive model for high-impact experiential learning. In Y. Huang (Ed.), *Handbook on credential innovation for inclusive pathways to professions* (pp. 418–434). IGI Global.

Acknowledgments

We would like to acknowledge the following partners, colleagues, and students who have contributed to specific chapters and sections of the book.

Gideon Adeniyi
Joanita Ayenyo
Sara Bramlage
Bikeke Saimon
Ashley Brito
Hart Hagan
Christopher Haines
Kennedy Mahili
Deo Mbabazi
Dr. Ndubueze Mbah
Muhammed Moinuddin
Mbilizi Kalombo
Ayomitomiwa Ogunsile
Hannah Ruth
Cynthia Tysick

The ELN Model and SDG Project Challenges were supported through State University of New York (SUNY) grants including Performance Improvement Fund (PIF) and Innovative Instruction Technology Grants (IITG).

Introduction

This book is about the transformative potential of Experiential Learning; both for students, and the world. While we often assume that student growth needs to happen first – namely, building skills and capacity through education and experience before making a contribution to the world, this book suggests the reverse. If we provide students with opportunities to add value to challenged communities and ecosystems through Experiential Learning, they will support broader impacts while building their own capacity.

This premise is more than an idea. Since 2019, we have been engaging students from the University at Buffalo (UB) in virtual Project Challenges featuring community-based organizations that are doing the work of sustainable development in West and East Africa. We support student engagement through our Experiential Learning Network (ELN) model, a home-grown system that includes three interconnected components: (1) a Project Portal that functions as a dynamic web-based market for creating and accessing mentored project opportunities; (2) PEARL modules that move students through Preparation, Engagement and Adding value, Reflection, and Leveraging their experiences toward broader impacts; and (3) digital badges that showcase achievement and help build value around engagement. Together, these components function as a dynamic system, transforming global sustainability challenges and collaboration with community partners into opportunities for individual and collective impact.

While the ELN model is admittedly bold, it is equally practical. Together the three components – Project Portal, PEARL, and digital badges, support the efficient administration and assessment of

Experiential Learning at scale. The model can accommodate virtually any type of applied learning activity as long as it is mentored, collaborative, and results in a tangible outcome. By adopting these minimal requirements, we can accommodate internships, research, study abroad, or other existing offerings, while supporting the creation of new opportunities and experiences to meet growing demand.

When we designed the ELN model we had practical considerations in mind. As a R1 public research university, our institution offered extensive opportunities for engagement through academic departments and centrally administered programs. But access was uneven and navigation challenging, especially for students who were uncomfortable approaching faculty or unsure of their specific goals or interests. By creating a searchable portal, we hoped to make Experiential Learning more accessible while also supporting students throughout the life of their project. Through guided facilitation administered through online PEARL activities, we could help students unpack and integrate their experiences with academic and professional goals and showcase their accomplishments for key audiences including employers and graduate schools. Through this system, we could also assess growth and movement through PEARL activities, utilizing data to strengthen the model and build further capacity for engagement.

When we launched the ELN model in 2019, there were 90 projects in the portal, 89 of which were mentored research experiences. The 90th project was an opportunity to engage with community partners in Tanzania, East Africa, through virtual collaboration related to a yearly study abroad course. The course brought together students from diverse majors and programs of study and featured community partners who were doing the work of sustainable development and climate action. Students were challenged to add value through customized projects that connected with their own goals and interests while addressing the needs and priorities of the featured partners. In designing the ELN model, we made sure that the system could accommodate community-based partners, including those featured in study abroad courses, with the goal of furthering collaboration between trips, and building generative impacts through customized projects.

To be clear, the value of engaging students with community partners around sustainable development was evident before the ELN model was created. We knew that students were eager to explore places and cultures, to engage with organizations working to support the most vulnerable communities and ecosystems. And we had observed, first hand, the transformation that happens when students are compelled to make a difference, to add value through their own engagement and collaboration. While our Tanzanian partners were special, we knew that there were many other organizations doing the work of sustainable development around the world, and eager to collaborate and leverage engagement to support their communities. But while the potential was clear from the beginning, it wasn't fully actualized until the need for virtual opportunities became urgent.

When the pandemic hit in 2020, our university campuses closed and students were sent home to finish their classes online. Research laboratories shut down, internships and clinical placements were canceled, and study abroad trips were disrupted. While students waited for life to return to normal, they began to search for Experiential Learning opportunities, visiting our online Project Portal and contacting the ELN for help. While our model was entirely web-based and able to support students from anywhere in the world, the majority of our project opportunities featured in-person engagement, and as a result were temporarily unavailable. So we began to add projects that we were able to facilitate and mentor ourselves, engaging our own relationships and networks. We began to add additional global partners who were doing the work of sustainable development and were eager to engage with students through collaborative virtual projects.

As demand continued to grow, we began to experiment, deepening engagement with the Sustainable Development Goals (SDGs) and connecting our model with academic programs and courses. With the extended disruption to credit-based Experiential Learning, there was an eagerness to embrace the functionalities provided by our system and the opportunity to connect students and classes with global partners. We found that the same technology-supported platforms that were being embraced for online instruction worked equally well for engaging our partners, even those in remote areas. Amidst the disruption to study abroad and international travel, we created a virtual version of our Tanzania study abroad trip, and worked with colleagues from SUNY COIL (Collaborative Online International Learning) to develop a Global Commons experience that featured Open Education Resources focused on the SDGs engaging students from across the State University of New York system in mentored projects with community partners. And we worked with our own faculty to integrate projects and engagement within diverse course structures and formats, allowing students to access meaningful experiences while fulfilling curricular requirements.

When the pandemic eased and students returned to campus, we continued to push forward, tightening our model while deepening engagement with partners around our shared work of sustainable development. As we write this book, the potential for our approach is more compelling than ever. The challenges related to climate change are accelerating and becoming increasingly complex, calling for innovative models that can translate within local contexts and communities. And young people everywhere are eager to access experiences that can help them develop skills and competencies and connect with potential jobs. Yet amidst the great need and potential, there is also growing danger. As higher education struggles with its own sustainability within increasingly challenging landscapes, many institutions are shifting their focus inward, with visions and goals becoming increasingly narrow and constraining, threatening the very innovation that can offer new opportunities and relevance.

Through this book and the sharing of related impact stories, we hope to convince readers that there is no better time to be bold with Experiential Learning, especially when boldness can support our own practical needs and sustainability. Students need Experiential Learning, not just any experiences, but those that are compelling and meaningful; experiences that demonstrate an ability to add value and address the problems we have created. Climate action is the ultimate laboratory for Experiential Learning and community partners around the world are poised to collaborate and lead the work, allowing educators and students to engage in ways that are both appropriate and transformative. This is the promise of Experiential Learning.

Part I

Expecting More from Experiential Learning

1 A Case for Project Challenges

This book begins with a simple yet compelling premise; that for today's students, coursework and grades are no longer enough. Beyond learning, young people need evidence of doing and experiences that resonate with key audiences including employers, graduate schools, and other keepers of resources and opportunities. Increasingly, high-value experiences are the coin of the realm, precious to those who seek them but also for communities and ecosystems that stand to benefit from their development. Since skills and competencies are best cultivated through real-world challenges, our most urgent problems can become our most powerful opportunities, if supported and integrated with academic and professional preparation.

This is the promise of Experiential Learning, a dynamic category of applied experiences becoming increasingly popular in American schools, yet largely nonexistent throughout much of the world. Most often associated with internships, mentored research, study abroad, or civic engagement, Experiential Learning is spawning new methodologies and approaches, connecting students with people, places, ideas and opportunities. It is being offered in-person and remotely, for-credit and informally, by faculty, teachers, and practitioners through diverse platforms and formats. While exciting from the standpoint of flexibility and customization, the sheer openness of Experiential Learning can be overwhelming, leaving many unsure how to meet the moment. This uncertainty is creating new markets for external vendors and purveyors of experiences, and missed opportunities for education and educators to innovate and lead the way.

Higher education already recognizes the value of Experiential Learning and the need to make it accessible for all students across

DOI: 10.4324/9781003489337-2

degrees and programs of study. The pedagogical benefits of Experiential Learning are compelling and well supported by research on High-Impact Practices (HIPs), a select group of experiential approaches associated with pronounced gains in student engagement, persistence, and retention, and enhanced benefits for students from historically underserved populations (Kuh, 2008). High-Impact Practices include mentored research, study abroad, project-based learning, and other approaches that feature significant time and effort, real world applications, reflective opportunities, and constructive feedback from skilled teachers or mentors. In addition to boosting student success and retention, these practices, when done well, cultivate competencies related to critical thinking, civic engagement, ethical reasoning, and global learning; competencies that offer societal benefits in addition to supporting student growth and capacity building (Kuh and O'Donnell, 2013).

What does it mean to do Experiential Learning well? It means approaching Experiential Learning as a teaching and learning paradigm, rather than simply connecting students with opportunities for external engagement. This contrasts with the growing trend of requiring students to find their own Experiential Learning opportunities, or leaving supervision and mentoring to external partners with limited integration of experiences with coursework or prior learning. While these types of offerings may satisfy growing demand and related resource challenges, they do not fulfill the promise of Experiential Learning or support the process through which students achieve transformational benefits.

Research on Experiential Learning Theory (Kolb, 1984) has demonstrated that while students drive the process of active learning, facilitators in the form of teachers and mentors are critical to supporting their movement through the dynamic process. To transform experiences into the creation of new knowledge, which Kolb suggests is the ultimate goal of Experiential Learning, students must work through a number of stages including concrete engagement, reflective observation, abstract conceptualization, and active experimentation, not necessarily in that order (Kolb, 1984). The more students give to this work of engaging, reflecting, and integrating their experiences, the more they get in the form of growth and capacity building. And the more they grow and build their individual capacity, the more potential they create for broader impact.

This dynamic relationship between giving and getting is critical to understanding and leveraging the promise of Experiential Learning as a catalyst for societal benefits. This relationship is most easily illustrated through the lens of workforce development and growing interest in internships and career readiness. As students engage in internships and high-impact work experiences, they develop 21st century skills and competencies that align with employer needs and priorities. Research suggests that in addition to professional success, these competencies can support economic development through increased productivity and a more seamless transition from school to work (De Fruyt, Wille, and John, 2015). In this way, internships and work experiences can simultaneously support students, employers, and the broader economy, catalyzing both individual and collective impacts.

While focus on employment and job success will continue to grow along with demand for internships and work experiences, there are other sets of competences that can catalyze both individual and collective benefits. These include skills and competences related to innovation, future thinking, sustainable development, and other emerging frameworks that prioritize problem solving, nimbleness, and adaptation, and leveraging technology to solve complex challenges. The work of cultivating these skills and competencies for the benefit of students, society, and the planet calls for experiences that are sufficiently powerful and aligned with the outputs we desire. Stated another way, in order to cultivate problem solvers, we need sufficiently powerful problems to solve.

This sets up an exciting hypothesis; that our most compelling societal challenges can generate powerful opportunities for students to develop high-value skills and competencies along with broader impacts. This suggested relationship has already been supported by research on problem-based learning (PBL), a High-Impact Practice that engages students in ill-defined problems toward producing tangible solutions or outputs. As students work through innovation and problem solving, they are supported by mentors and related resources and expertise. PBL, when successfully supported, is associated with deep learning and competencies that are highly transferable and long-lasting (Potvin et al., 2022). In addition to the actual products of engagement, which can be valuable in their own right, PBL can also support agency and civic-mindedness along with valuable 21st century skills.

Research on the effectiveness of PBL suggests that associated impacts can be further boosted by connecting students with individuals or communities that are directly impacted by the problem being addressed (Miller and Krajcik, 2019). By engaging with people who are closest to both problems and potential solutions, students can access opportunities to cultivate understanding, empathy, and insights which in turn can support motivation and creativity. In fact, the more students are compelled to help and add value, the more they are willing to stretch their own learning, developing new skills, competencies and understanding. Engaging with affected communities and partners also helps to establish local context, which is often missing in the analysis of complex problems and the development of effective solutions.

PBL is already being utilized to address a range of societal challenges including humanitarian, health-related and environmental disparities. Through competitions, summits, and related events, students, often in teams, are connected with experts and community representatives and given access to digital resources related to target challenges. While powerful, these opportunities are often highly selective with competitive application and selection processes. To minimize the demands on facilitators and teachers, PBL is often limited to high-achieving students who already possess foundational skills and knowledge, and the capacity to self-direct and leverage available resources. While exciting for participants, there is often little opportunity to build on broader impacts or leverage PBL for the students, communities and ecosystems that stand to benefit most. As a result, this approach remains largely untapped and underutilized.

What would be needed to activate the potential of PBL to catalyze both individual and collective impacts at scale, to help students contribute to climate action while supporting their own academic and professional growth? If we recognize the potential of environmental and humanitarian challenges to provide project opportunities for students, we could transform our schools and education programs into engines for sustainable development and innovation. Teachers and professors could be mentors and content experts, and community partners could offer pathways to contribute and add value through engagement and collaboration.

When we see this potential, we recognize the importance of crafting Project Challenges for maximal resonance with students and the communities and ecosystems that stand to benefit from their activation. What constitutes an effective Project Challenge? As in all aspects of

Experiential Learning, there is room for experimentation and creativity. But to be effective, projects must provide a mechanism to connect engagement with the communities and ecosystems that stand to benefit, and be sufficiently motivating to keep students engaged, giving the best that they have to offer. If we can design Project Challenges that are inherently powerful and ready for activation, we can build dynamic systems that transform Experiential Learning into a catalyst for sustainable development.

Project Challenges Up-Close: Water Recovery

Written with Ayomitomiwa Ogunsile, Founder, Ayomi Arts, Nigeria

The tie-dye fabric, popularly known as adire in Nigeria is one of the cultural symbols of the southwestern people but has spread to other regions of the country including Kano in the north. The name adire was first applied to indigo-dyed cloth decorated with resist patterns in the early 20th century. Resist dyeing is a traditional textile dyeing technique used to create patterns and designs on fabric by selectively preventing the dye from reaching certain areas. More modern techniques include wax resist, cassava paste resist, and other methods used to achieve beautiful and vibrant designs using both natural and synthetic dyes.

With the embrace of new designs, textiles, and technologies, the market for adire has grown to include both local and international buyers, and features homewares, clothing, arts and other emerging categories of products. Although adire is still practiced predominantly by women (70% of current producers), it has become a cultural symbol of Nigerian artists but also a commercial practice that contributes to both the formal and informal economies. Since unemployment in Nigeria is high, especially among young people and women, adire and textile fabrics represent an important way to preserve and celebrate tradition and heritage while generating revenue and supporting livelihood. In this way, adire represents an exciting lever for cultural and economic growth.

Yet behind the beautifully made adire fabric is the ugly reality of its production cost. Through the practices of adire, contaminated wastewater is generated and dumped into Nigerian water bodies and open drainage systems. The chemicals used in the dyeing process, when released untreated, pollute the environment causing health problems for the artisans and impacted communities. These include skin rashes, ulceration, swelling, respiratory diseases, and complications during delivery. Water contamination disturbs aquatic and land biodiversity, creating problems that exacerbate the effects of climate change and impact the most vulnerable. In this way, adire can be viewed as both an economic asset and an environmental threat. But it is also an important opportunity for innovation and problem solving. How can Nigerians preserve adire and leverage its potential value while also preserving the cleanliness of our precious water? How can we rethink how wastewater from adire production can be managed sustainably?

We invite students and educators to help design a system to recycle wastewater from the textile dyeing process. We are seeking solutions that are easily adaptable for small and medium-scale producers and scalable for larger enterprises; processes that are eco-friendly and affordable. We require scalable systems that include elimination of toxic chemicals in the environment and water bodies, reduction of water stress by channeling back recovered water into the production process, reduced dependence on freshwater resources, improved production efficiency, and solutions that are affordable and accessible to cottage producers yet scalable for bigger enterprises.

The Wastewater Recovery Project Challenge is open to students and educators of all backgrounds and grade levels around the world who share a commitment to sustainability and innovation. We are ready to test your ideas through collaboration and experimentation. While all students and educators are welcomed, we are especially interested in engaging Nigerian and African students and educators through in-person and virtual Experiential Learning.

References

De Fruyt, F., Wille, B., and John, O. (2015). Employability in the 21st century: Complex (interactive) problem solving and other essential skills. *Industrial and Organizational Psychology* 8(02), 276–281.

Kolb, D. A. (1984). *Experiential learning: Experience as the source of learning and development* (Vol. 1). Englewood Cliffs, NJ: Prentice-Hall.
Kuh, G. D. (2008). *High-impact educational practices: what they are, who has access to them, and why they matter*. Report from the Association of American Colleges and Universities.
Kuh, G. D., and Ken O'Donnell. (2013). Ensuring quality and taking high-impact practices to scale. *Peer Review* 15(2), spring, 32.
Miller, E. C., and Krajcik, J. S. (2019). Promoting deep learning through project based learning: A design problem. *Disciplinary and Interdisciplinary Science Education Research* 1, 7.
Potvin, A. S. et al. (2022). Mapping enabling conditions for high-quality PBL: A collaboratory approach. *Education Sciences* 12(3), 222.

2 Activating the Sustainable Development Goals (SDGs)

The idea of inviting young people to address sustainability challenges is more than an educational exercise. It is an essential component of climate action, one that aligns closely with the Sustainable Development Goals (SDGs). Born from the United Nations Conference on Sustainable Development, held in Rio de Janeiro, Brazil, in 2012, the SDGs are a set of 17 global goals related to environmental, political, and economic challenges that are critical to preserving humanity and saving the planet. Together, they provide a roadmap to sustainability.

While the SDGs may be unfamiliar by name, they are often recognized by their visual representation – 17 brightly colored squares with the name of each goal boldly displayed along with its corresponding number. Stacked in three rows, resembling colorful building blocks, the SDGs are easy to identify and explore. They include:

1. No Poverty;
2. Zero Hunger;
3. Good Health and Well-Being;
4. Quality Education;
5. Gender Equality;
6. Clean Water and Sanitation;
7. Affordable and Clean Energy;
8. Decent Work and Economic Growth;
9. Industry, Innovation and Infrastructure;
10. Reduced Inequalities;
11. Sustainable Cities and Communities;

DOI: 10.4324/9781003489337-3

12. Responsible Consumption and Production;
13. Climate Action;
14. Life Below Water;
15. Life on Land;
16. Peace, Justice and Strong Institutions;
17. Partnerships for the Goals.

Although presented as 17 discreet goals that can be considered individually, the SDGs are highly interconnected, comprising 169 targets and 247 indicators, of which 92 are environment related. Targets function as component goals that must be met both individually and collectively in order to be achieved, while indicators function as assessment metrics, allowing for tracking of corresponding progress. In this way, the SDGs are highly functional, designed to set a clear course for collective action, while supporting and encouraging customized responses and robust research and assessment.

While the SDGs have prioritized policymaking and national development planning as key drivers of action, they have also had an influence on economic development and innovation. By framing both challenges and outcomes in the form of goals, targets and metrics, the SDGs invite innovation, problem solving and the design of new technology-based solutions. They also provide opportunities for existing technologies to be translated and adapted to address needs and challenges within local contexts. Efforts to connect emerging innovations with the SDGs are resulting in important breakthrough technologies that are driving investment, economic development and cultivation of human capital. For example, there is growing interest in leveraging artificial intelligence (AI), machine learning, robotics and other computer-supported innovations to mitigate or even reverse the effects of climate change. Advancements in renewable energy, 3D printing, drones, and gene technology, all have humanitarian applications that can be developed for the greater good while also supporting economic growth and job creation. Scientists, researchers, and businesses are being incentivized to lead this work, embracing the SDGs as invitations to innovate and scale, creating new markets and opportunities for sustainable development.

This idea of using the SDGs as a springboard for innovation and problem solving is equally exciting for students and young people seeking meaningful Experiential Learning opportunities. In the same way the SDGs set out needs, solutions, and progress metrics for experienced scientists and researchers, they serve as invitations for young people to make contributions through project outputs, while exploring emerging fields and opportunities. If young people could engage with the SDGs, exploring technology-supported solutions through Project Challenges, they could support their educational and professional goals while cultivating valuable skills and competencies.

As pedagogical tools, the SDGs offer exciting benefits that make them especially suited for Experiential Learning. Because the 17 SDGs are presented as discreet yet highly interconnected goals, they support learners in exploring both the complexity of challenges and the elegance of scalable solutions. To understand this functionality it is helpful to think of SDGs as contextual frames. Grounded in theories of cognitive psychology, contextual frames support linkages between ideas and concepts, allowing learners to see similarities and draw inferences (Slominski et al., 2020). Contextual frames provide insights for understanding complexity across different modalities such as language learning or deciphering complex reading passages. Contextual frames allow individual ideas to be linked or bound together, helping learners track references as they appear in different contexts and places, gaining a deeper understanding of both similarities and nuanced differences. Contextual frames can be useful in interdisciplinary research, innovation, or other activities involving complex information and analysis.

How can students use the SDGs as contextual frames? Because the 17 goals are both discrete and interconnected, they can be used to generate insights and understanding across different areas of focus and analysis. For example, SDG 5 Gender Equality can be used as a contextual frame for exploring SDG 4 Quality Education. Once a student understands that girls often bear the burden of household chores such as fetching water and caring for younger siblings, they can bring this understanding to their exploration of education challenges including differential retention and graduation rates for girls in rural villages. While SDG 4 can help students understand the challenges associated with Quality Education, it can also support the formulation of scalable solutions, for example, making sure that education interventions address family and societal expectations that may limit consistent participation for girls. While teachers can introduce the SDGs in formal instruction, contextual frames are most powerful when generated by learners themselves. Accordingly, as students engage with Project Challenges, they can use the SDGs to support their own discovery, challenging their understanding and building higher-level skills and competencies as they work toward generating and implementing solutions.

The space between complex problems and scalable solutions represents a dynamic playground for Experiential Learning. By engaging students in sustainability-related Project Challenges, we can transform our most pressing problems into opportunities for innovation and skill development. But how do we move from exploration of the SDGs to climate action, as suggested by the title of this book? Students want to do more than explore, they want to take part in solutions, not just studying or proposing ideas but actually contributing in ways that are meaningful. Through connecting students with organizations that are doing the work of sustainability within local communities and ecosystems, we can activate their potential to contribute to climate solutions. Since climate action is inherently local, we can support the development of effective solutions while cultivating high-value skills and competencies associated with economic growth, innovation and sustainability.

SDGs Up Close: Climate Action

Written with Christopher A. Haines, Biodiversity for a Livable Climate

SDG 13 – Climate Action– urges us to combat climate change and minimize its disruptions. Its targets focus on strengthening resilience and adaptive capacity to climate-related hazards and natural disasters, planning and management, education, mitigation and capacity building measures. While critical, the targets included in SDG 13 are alone insufficient to address the growing threat and impacts of climate change (Haines, 2021). Nature and climate are incredibly complex systems that cannot be understood in their parts. Accordingly, it is necessary to consider SDG 13 Climate Action together with 6 Clean Water and Sanitation; 14 Life Below Water; and SDG 15 Life on Land.

We must recognize the far-reaching impacts of human-created disruption and the need for human-initiated action. Through deforestation, development, and extractive practices we have disrupted the ability of climate systems to self-regulate and provide for the needs of adjacent communities who rely heavily on nature for sustaining their livelihood. As a result of our actions, the same communities that are most dependent on the outputs of ecosystems, are the most vulnerable to the resulting climate crises which are exacerbated by further degradation and destabilization.

When we examine the underlying problems and levers of action associated with climate change, the restoration (or regeneration) of biodiversity emerges as especially powerful. Biological diversity, or biodiversity, refers to the variety of life on Earth, in all its forms, from genes and bacteria to entire ecosystems such as forests or coral reefs. Loss of biodiversity through deforestation, conventional agriculture, and other destructive human practices is destroying the Earth's mechanisms for achieving resilience and adapting to changing conditions and threats.

The United Nations asserts that biodiversity is our strongest natural defense against climate change. The UN explains that the Earth's land masses and oceans serve as natural carbon sinks, absorbing large amounts of greenhouse gas emissions and that conserving and restoring natural spaces and the biodiversity they contain is essential for limiting emissions and adapting to climate impacts.

Our earliest understanding of climate recognized that the atmosphere absorbs heat radiated from the sun-warmed surface. To understand how biodiversity interacts with these climate physics to be a key component of climate action, we must explore the properties of energy and heat. This begins with the first law of thermodynamics that states energy can change form but never completely vanishes. When solar energy strikes vegetation, the majority converts into chemical bonds that promote growth and transpire water, causing cooling.

When we transform green spaces through urbanization, the materials absorb and store the energy and reradiate it as heat. While there are multiple causes of urban heat islands, the number one reason we have overheated cities is the lack of vegetation. With hundreds of millions of acres of agricultural lands cleared to bare ground, deforested, aridified, or transformed by man-made materials, the impacts are not limited to cities.

In cities, the man-made materials used in construction are both incredible absorbers and radiators of thermal energy. This creates what scientists describe as an urban heat island (UHI).

With UHI, the stored energy continues to be reradiated for a longer period than a surface that is covered by vegetation. This results in warmer evening temperatures lasting longer, while limiting the amount of time when surfaces would cool.

The compounding impact of warmer than usual overnight lows serves to increase the initial temperature that the air heats from the following day. This can be a recursive process and in areas where a UHI effect is in place dramatic heat waves occur. Between 1800 and 2000, urban development increased more than 100 times faster than greenhouse gasses. Urbanization and the challenges that we face because of it are at the heart of SDG 11.

When we recognize the actual cause of this warming, we find a hopeful future. We can reverse the decisions that have led to this increased warming by restoring green space, creating linking corridors, installing living building surfaces, restoring air-flow and engaging around reforestation projects and initiatives. The true optimism lies in the fact that this can be done completely, locally, and rapidly, in months, years or maybe decades, not centuries.

References

Haines, A. (2021). Greenhouse gases: True, but not the whole truth. *Journal of Sustainability Education* 25, June 2021. ISSN: 2151–7452

Slominski, T., Fugleberg, A., Christensen, W. M., Buncher, J. B., and Momsen, J. L. (2020). Using framing as a lens to understand context effects on expert reasoning. *CBE Life Science Education* Sep; 19(3), ar48. doi: 10.1187/cbe.19-11-0230. PMID: 32870088; PMCID: PMC8711834.

3 Community-Based Partners

While the SDGs can serve as powerful contextual frames for Experiential Learning, they require the infusion of local engagement to activate transformative impacts. And since climate action happens at the local level, within the most vulnerable communities and ecosystems, it is critical to engage local communities, both in exploration of challenges and implementation of solutions. This creates an important opportunity for engagement of community partners in sustainability-related Project Challenges.

The term community partner, and partnerships, can mean many things in the worlds of education and sustainable development. Our model focuses specifically on community-based organizations (CBOs), non-government organizations (NGOs) and civil society organizations (CSOs) that are on the frontline, doing the work of sustainability within vulnerable communities and ecosystems. These locally-led groups focus on a diverse array of humanitarian and environmental challenges, offering unique programs and interventions to supplement resources provided by governments and external stakeholders. The number of community-based organizations around the world has continued to grow, although actual numbers are difficult to obtain.

While grassroots organizations have been viewed as critical partners for humanitarian and development work, their relationships with local governments can be contentious. Since many have played watchdog roles, holding governments and systems accountable while advocating for the most vulnerable, they are often left out of conversations and access to resources. Yet with the emergence of climate action as a unifying priority, community organizations are

DOI: 10.4324/9781003489337-4

increasingly being seen as critical partners and invited to the table. Because they are working within the most vulnerable communities and ecosystems and offer capacity in the form of organizational infrastructure, relationships, and key resources, they have much to contribute. And because climate change is exacerbating the vulnerability and needs of the communities they serve, many community-based organizations are adding environmental projects including tree planting, water filtration, and regenerative agriculture to their portfolios.

While the proliferation of community-based environmental projects offers important opportunities for climate action, there are fundamental capacity issues to be addressed. The community-based organizations leading these efforts are often lacking in resources and support, making them inherently fragile from an organizational standpoint. Program costs related to training, materials, equipment, and supplies, and assessment support are considerable, leaving community partners eager to find collaborators with related expertise and resources. As community-based organizations deepen their focus on climate action, they will require deeper engagement from stakeholders with aligned resources and goals, including higher education and resources offered through the work of Experiential Learning.

For students looking to develop their own capacity through Experiential Learning, community partners offer valuable opportunities for growth and competency development. To appreciate this value, it is useful to consider a concrete example. Imagine a student completing a Project Challenge focused on regenerative agriculture practices in Sub-Saharan Africa. Through exploring digital resources, the student could evaluate various methodologies being implemented by practitioners in the field. They could deepen their engagement by conducting hands-on activities and experiments, allowing them to transform their learning into doing, under the mentorship of teachers or practitioners. This potential is already exciting and suggests the value of using the SDGs to support student discovery and Experiential Learning. But imagine if the same student were connected with an actual community partner using a specific regenerative agriculture practice to mitigate the impacts of climate change within a particular region; for example, farmer-managed natural regeneration (FMNR) in rural Tanzania.

Known informally as "stump regeneration," FMNR supports the regeneration and management of trees and shrubs that sprout from stumps, roots, and seeds. This method is being used in areas that have been degraded and deforested by cutting down trees for firewood and also burning for farming purposes. Through engaging with an organization that is implementing this practice, the student could go much deeper with their exploration. They could examine the context and gain a personal understanding of the effects of deforestation and the variables that contribute to its practice, the impacts on people's lives, and how FMNR is emerging as a promising solution. And as the student engages and explores the intervention, they could identify ways to make meaningful contributions through project outputs and leveraging technology-based resources available to support their engagement.

What types of project outcomes could be meaningful to community-based partners and the work of climate action? It is important to recognize that virtually any project deliverable can be of value if aligned with the needs of the partner and their efforts to support sustainable development. This includes the types of deliverables that students are eager to share with potential employers or graduate programs. Research briefs, educational materials, marketing and digital media, data maps or grant proposals can all be valuable to organizations struggling to build capacity and connect with resources. In addition to these types of project outputs, community-based organizations need access to web-based platforms and knowledge of how to effectively use them. When we imagine Project Challenges designed to leverage technology-based resources, the possibilities for impacts become increasingly exciting. By virtue of their affiliations, students have access to powerful tools that can support local interventions and contribute to collective amplification and progress. In fact, the very technologies that are driving sustainable innovation, such as AI, satellite mapping, drones, and machine learning, are being developed and tested within the same colleges and universities that are tasked with supporting Experiential Learning.

While the potential for technology-supported innovation is exciting, we should not dismiss the value of direct hands-on experiences that engage students in the work of climate action. With swelling numbers of students, recent graduates, and out-of-school youth across the world, there is both massive potential and need for experiences that cultivate skills and competencies through meaningful engagement. By connecting youth with community-based partners doing the work of sustainable development within nearby regions and ecosystems, what could we achieve and how could students benefit from their engagement?

When we recognize the potential of community-based partners as critical players in activating the power of Project Challenges and Experiential Learning, it is difficult to imagine doing this work without them. But recognizing this importance creates an additional responsibility; namely, ensuring that we can add value to their work, helping them build capacity while retaining their ownership and control. Rather than pushing in our own solutions or interventions, we must play a supporting role, framing opportunities to collaborate and add value, and positioning our partners as innovators and stewards of sustainable solutions. Once we recognize that the work belongs to those who are closest to both the challenges and potential solutions, we will see exciting opportunities and spaces to engage, deepening our own commitments and investments in Experiential Learning and climate action.

Community-Based Partners Up-Close: Climate Action Journalists

Written with Bikeke Saimon, Founding Coordinator of Umbrella for Journalists in Kasese (UJK)

When it comes to sustainability-related Project Challenges, potential partners are everywhere. In identifying possible collaborators, it is helpful to consider their ability to help students explore challenges while also contributing to solutions. When we apply this filter, climate action journalists emerge as powerful partners. Climate action journalists are working in vulnerable communities and ecosystems around the world, capturing and sharing stories towards the goal of catalyzing change. These journalists, often young people directly from the impacted communities, are getting close to crises and catastrophes and the people most affected, while also telling stories of hope and resilience. They do so with the intention of transforming stories into action, mobilizing audiences and stakeholders to drive positive change.

Bikeke Saimon is a Ugandan journalist and founding coordinator of Umbrella for Journalists in Kasese (UJK), which promotes freedom of expression, safety for journalists, content curation and the amplification of community voices. UJK also skills and nurtures a new breed of media practitioners and content curators on environmental reporting, with the purpose of changing the narrative around environmental breakdown and becoming part of the solution. Bikeke views climate stories as invitations for climate action and seeks to empower local youth to both share their own stories and become agents of change. He suggests that many people, including youth in developing countries, do not understand how climate change is impacting communities including his own in Uganda. And young people in impacted communities lack a sense of urgency or belief that they can make a difference. He worries that a growing acceptance of climate change is preventing communities from taking action and implementing the practices that can turn things around. By telling stories, and encouraging youth to share their own, he sees climate action journalism as a portal for impacts and environmental regeneration.

Bikeke offers a recent article as an example of how climate action journalism can engage students from around the world in exploring climate change and sustainable development within local contexts, while identifying opportunities for intervention and action.

Contaminated water from the Nyamwamba valley which hosts Kilembe mines industry in Kasese District is exposing the vulnerable communities to metal poisoning.

The valley is the bed for river Nyamwamba, a major water source for communities in the low land areas of Kasese and the chief water supply that feeds into the historic Mubuki irrigation scheme.

However, a recent strange color and strong stench covering the river has triggered fear of contamination among residents suspecting it could be polluted with copper and cobalt minerals.

The residents suspect that the contaminants are deposits of minerals left unprocessed and copper waste deposited along the river, getting eroded into the river water since the heavy floods of 2020.

There is already fear that foods and forage grown around this area most likely contain significantly higher concentrations of copper, cobalt, zinc and in turn have been consumed by the local people.

The consumption of food contaminated with these elements can cause cancer, gastro-intestinal complications, a decrease of immunological defenses, physical and mental disabilities and retardation.

In August 2019, during the cancer run activities in Kasese town, Kasese was highlighted for having contributed 50 percent of cancer patients at the Cancer institute Nsambya.

Yeresi Biira, a palliative care nurse specialist with Cancer and Aids Relief Organisation (CARO) said that if the accumulation of cancer cases is not arrested today, by 2025 many people from the region could be diagnosed with the disease.

A social survey conducted by Makerere University students established that more than half of the households in Kilembe (51%) depended on tap water, 38% on the river Nyamwamba while 11% collected water from community water sources such as streams, water wells and gravity water systems.

The survey also highlighted that the presence of metal sulfide deposits within an ecosystem can have significant impacts on the quality of water and on the health of humans.

Speaking to UJK, Solomon Kule an environmental specialist within Nyamwamba Division pointed out that indeed there are clear signals that the water in the river is contaminated.

Apart from change of color, Kule says food crops and vegetables about 100 meters off the river bed are beginning to change color or drying-up.

Kule advises farmers to grow food crops on the upstream or hills.

Kenneth Muhindo a youth in Kasese town said that the lack of alternative sources of water forces residents to drink the contaminated water resulting in suffering from a range of waterborne diseases.

He quickly calls for massive intervention from the government to have stand-up taps expanded in communities.

The Kasese Municipal agricultural and production officer Sanairi Bukanywa agrees that Nyamwamba water is contaminated but hastens to add that the Government is taking steps to protect the water source from being contaminated.

Water in tributaries flowing into River Nyamwamba, located outside the mine and tailing zone also contain elevated concentrations of heavy metals, exceeding recommended drinking water thresholds for Fe (38%), Ni and Mn (13%).

Heavy metals and trace elements found upstream of River Nyamwamba and in the tributaries possibly originate from geological weathering, attributed to area mineralogy and geology.

Whereas River Nyamwamba is well known for the destruction of property and loss of life whenever it overflows, the environmental effects are less known and could be even more lethal.

The role of clean water in alleviating poverty cannot be overstated. While stories such as this one can introduce environmental problems contributing to specific SDGs, they can also lead to solutions and innovations that can be piloted and tested within communities. Bikeke envisions young people being engaged in all of it, from identifying problems, to innovating and implementing solutions, and sharing stories of hope and action. As interest in Experiential Learning and Project Challenges grow, he hopes climate action journalists will be invited as valued collaborators. In the meantime, he is galvanizing fellow journalists to be ready for the opportunity.

4 Mentors, Teachers, and Content Experts

When we think about mentoring with regard to Experiential Learning, we often assume extensive supervision and oversight, usually by faculty or course instructors. These expectations can be daunting, causing a reluctance to take on mentees or to integrate Experiential Learning within courses or programs of study. While high-touch mentoring can be transformative for students, it is not the only way for faculty and teachers to support Project Challenges. If we are serious about developing these opportunities and the benefits they afford, we need to clarify what is critical in terms of facilitation and support, and consider how to best offer it at scale.

As detailed in earlier chapters, the Experiential Learning cycle offers several important roles for facilitation and mentoring. These include structuring engagement opportunities, supporting perseverance and follow-through, providing access to specialized knowledge and expertise, and helping students integrate experiences with prior knowledge, academic learning, and career goals. While it is perhaps possible for a single person to provide all of these types of support, it is more practical to examine each need individually. By doing so, we can design ways to offer effective support across diverse contexts, environments and scales.

By recognizing the importance of helping students move through the various stages of Experiential Learning, and sticking with a project until outcomes are met, we can provide facilitative support and monitoring of student progress. While this type of facilitation is critical to activating the potential of Experiential Learning at scale, it is possible, and even advisable, to provide it through technology-supported activities rather than through teachers, faculty or mentors.

DOI: 10.4324/9781003489337-5

In Chapter 6, we detail the PEARL engagement framework which provides a flexible process for moving students through Project Challenges as they engage with mentors and partners. PEARL stands for Prepare; Engage and Add value; Reflect; and Leverage. We have translated PEARL into online activities that help students get the most of their experience while also supporting dynamic research and assessment. By providing facilitation through online activities, we can release expectations of teachers and mentors to go beyond what they do best, allowing them to focus on areas of synergy and alignment.

How can teachers connect with Project Challenges without having to support movement through the various stages of Experiential Learning? With PEARL in place, teachers can focus on helping students connect experiences with coursework and academic learning. By integrating Project Challenges and PEARL activities within courses and programs of study, teachers can leverage engagement to enhance instruction and support academic goals. In Chapter 8, we present examples of Project Challenges that have been integrated within diverse course formats including study abroad, independent studies, and studio courses. By making PEARL activities universal and providing centralized facilitation, we can focus resources on helping faculty integrate Project Challenges within course syllabi and assignment structures. Faculty who leverage Project Challenges enjoy enhanced engagement and student motivation, which translate into course performance and broader impacts related to enrollment and retention.

Once PEARL and course integration are in place, opportunities for specialized expertise, technology-based tools, and other support can be offered to enhance student growth and related project outputs. As students become immersed in Project Challenges, getting close to the work and needs of partners, they are motivated to seek additional information, skills and tools that can be obtained through leveraging available resources including people, organizations, and technology. When students explore their programs and institutions through the lens of resources and expertise, they begin to recognize the bounty that education offers, and the potential value for their partners and the communities and ecosystems that stand to benefit.

While potential teachers and mentors are abundant within higher education, they need to be connected with Project Challenges in ways that are inherently incentivizing and supportive of core needs and priorities. It is useful to think of those contributing to and supporting Project Challenges as engaging in their own Experiential Learning, working towards outcomes with both individual and collective impacts. Like students, mentors must be motivated to sustain their engagement and stretch their own contributions to maximize outputs. Also like students, the more they give, the more they can receive in terms of growth and capacity building. In Chapter 8 we share ways that teachers and mentors can be supported and incentivized, connecting their engagement with key priorities related to teaching, research, grant development, and assessment while providing sufficient support and facilitation of their efforts. The work of incentivizing and supporting teachers and mentors is highly contextualized and must be thoughtfully designed to align with particular environments and cultures. Since teachers and mentors are foundational to the success of the model, it is critical to ensure that their engagement is inherently satisfying in addition to being meaningful for students.

Once teachers and mentors are in place, we have everything we need to transform Project Challenges into catalysts for climate action. Environmental challenges are everywhere and community partners are eager to engage. Students are seeking high-impact experiences and evidence of adding value and making contributions through project outputs. And courses and programs of study are ready to be enhanced through the addition of projects and engagement with community partners. When we view education through the lens of Project Challenges, we see a universe of possibilities and an urgent need for models that can connect students and faculty with community partners, leveraging PEARL to support individual and collective impacts.

Mentors Up Close: Digital Librarians

Written with Cynthia Tysick, Innovative Pedagogy and Creative Spaces Librarian, University at Buffalo

In the new world of Project Challenges, much of the work is data-driven and digital, making librarians powerful mentors and resources. From exploring the SDGs, to the work of partners and local solutions, it is all about contextualizing and brainstorming. Librarians can help find background or introductory information, consult on the scope and directions of projects, and provide training on effectively searching, evaluating, and exploring digital resources, including those that are free, open, or subscription-based.

The work of the librarian has evolved from the person in a physical library, reading books, and answering questions to someone who is trained in all forms of innovative technologies for finding, creating, and disseminating knowledge. Today's librarians work simultaneously in multiple media and are uniquely situated to support the development of digital literacy skills. They can help students determine which research is relevant to their project, to be critical of where information comes from, and to create compelling ways to digitally disseminate related work.

Cynthia Tysick is a digital librarian with a background in anthropology and a curiosity for all things tech. She helps students, faculty and partners get the most from their project experience. Cynthia's responses to the questions below illustrate the potential of librarians to support both the development and activation of Project Challenges, working with students, teachers and partners to catalyze both individual and collective impacts.

How do you support students as they work on Project Challenges?

I meet with students when they first get started, discussing why they are interested in the particular challenge and what resonated with them. It is similar to a research interview that I would typically do with a student writing a paper. I try to ask them about their own background, their major, or their hobbies. I find that if I am able to connect with something relevant, they are more apt to stay with the project, not only completing it but also adding value and getting an intrinsic reward. I have a large white board in my office and I begin drawing out some of the elements of the project, mapping out where their interests might intersect with the partner's needs. Often, students will take a picture of the whiteboard and think about it as they decide where they might want to engage and add value. If they already know, I show them how to do a background search to get contextual data, narratives, articles, chapters, or reports. We come up with some key terms to use in their searching so they can continue on their own. They are always encouraged to come back and see me to refine their information needs and gather more data as they work with their partner.

How do you engage with partners?
Our partners come from Low-Middle Income Countries (LMICs) and they lack access to high speed internet, quality research, and time to search. I see myself as their librarian and try to give them answers to their own information needs. I've often provided partners with articles, data, reports, and grant calls relevant to their work. They don't have the time to go through databases or open access websites to find what they need. Sometimes, once I have built a rapport with them, they will send me WhatsApp links to outrageous "fake news" and ask me if it's real. I've been their vetter of all things truth and I'm fine with that. I've also noticed that often they lack the capacity for themselves and their staff to run an efficient and effective organization. So, being the tech savvy librarian that I am, I started an online academy for them where I teach them to write business plans, search for funding, create compelling mission and vision statements, and use social media to their maximum benefit. When it comes to the projects they've developed, I discuss scope and setting realistic expectations. The visionary part comes from Dr. Huber when she and partners dream big. I know the skills and interests of students so I try to help the partner turn those dreams into multiple, achievable projects that our students will enjoy and be capable of doing. Some partners want funding so they ask for projects around fundraising. I try to dissuade them from that because a student's role isn't to generate funds. However, I encourage the partner to think about marketing or media materials around fundraising initiatives they run that our students could make. Students will always get a return on investment when they learn to create communication materials.

How do you connect the SDGs?
The Sustainable Development Goals are really research topics that students can explore throughout their programs of study. The SDGs examine the most pressing issues of their time and all seventeen are necessary to survive. A humanities student can see themselves in gender equality, peace, justice and strong institutions. Social science students gravitate towards decent

work and economic growth, gender equality, and zero hunger. Architecture and engineering students work within sustainable cities and communities and clean water and sanitation. For me personally, I connect with the often overlooked SDG 17, Partnerships for the Goals. Librarians see connections, facilitate the exchange of ideas, and have a global citizen mindset. We live to help people place themselves within their challenges and utilize the physical and digital landscape to succeed, in whatever form that may take. Librarians are happiest when those we help confidently harness their new digital literacy and digital media skills to create innovative solutions and collaborations.

Part II

A Scalable Model

5 Creating a Project Portal

The ELN model begins with the creation of a Project Portal, a digital space for showcasing available opportunities and encouraging the creation of new offerings. It is helpful to think of a Project Portal as a dynamic web-based market designed to attract students and project mentors, and incentivize engagement through the offering of strategic resources and support. Like any market, a Project Portal is designed with specific audiences and conditions in mind. And the ability to clarify and assess these variables is critical to setting up the model for success.

When we designed our Project Portal in 2018, we had several key goals that guided our process (Huber et al., 2021). These included: (1) ensuring that all students could find a meaningful project opportunity; (2) prioritizing access and equity; and (3) driving ongoing engagement in the model. These goals were framed within our unique strengths and challenges at the time. As a major public research university, we had a long history of engaging undergraduate students in mentored research and other Experiential Learning activities. But access was uneven across schools and departments, and many students had difficulty navigating offerings and expectations. On a campus of our size (roughly 21,000 undergraduates and 11,000 graduate students), it was often challenging, and intimidating, for students to make initial connections with professors. The Project Portal would help ease these challenges by expanding access to Experiential Learning, while helping students connect their engagement with coursework and programs of study. While our model would focus on students, we also needed to support faculty and departments to ensure a robust portfolio of opportunities and mentors.

DOI: 10.4324/9781003489337-7

Before starting our initial design work, we conducted an environmental scan of the current opportunities offered through the university. We found that although students indeed had access to diverse activities and resources, there were considerable obstacles that prevented many from accessing the benefits of engagement. Namely, the offices that administered internships, service-learning, study abroad, and creative activities, had individual rules and constraints, dictating what counted as Experiential Learning and what was required of participating students. In creating our ELN model, we would adopt an inclusive approach to Experiential Learning, with minimal requirements towards fostering collaboration with both internal and external partners.

To be featured in our Project Portal, experiences would need to be mentored – by faculty or staff at the university, collaborative – with at least three points of contact; and culminate in some tangible outcome meaningful to an identified audience. While additional conditions could be introduced for signature project offerings, including SDG Project Challenges, the minimal requirements would hold across all opportunities. Together, these requirements would provide assurance that all experiences would be sufficiently robust to produce meaningful outputs in the form of posters. presentations, performances, reports or analyses, or virtually any deliverable or output with an associated audience or purpose.

While our requirements would ensure meaningful opportunities for students, they would also accommodate existing projects and mentors, while supporting the design and testing of new offerings. To be successful, Project Challenges would need to attract students and resonate with priority audiences such as employers or graduate schools. With our minimal requirements as guides, the Project Portal could serve as an engine for innovation and experimentation. This generativity would be necessary to ensure a sufficient supply of projects while also supporting responsiveness to emerging areas of innovation and opportunity.

Since launching our Project Portal in 2019, we have strived to offer a variety of project options while honoring a commitment to mentored research, the mainstay of our university. However, this book is specifically focused on our SDG Project Challenges. These are signature offerings that connect students with partners who are doing the work of sustainability within communities and ecosystems around the world, with a focus on West and East Africa. This geographic focus originated from a study abroad trip to Tanzania and a mentoring affiliation with the Young African Leaders Initiative (YALI), and has since grown and expanded to include a diverse group of collaborators detailed in Chapter 13.

In addition to the universal project requirements of mentorship, collaboration, and tangible outcomes, Project Challenges have additional requirements of connecting with the SDGs, engaging students directly with featured partners, and working on local solutions to global challenges. Like all ELN projects, SDG Project Challenges are included in the Project Portal and are searchable by featured tags, filters, and keywords. The dynamic structure of our Project Portal allows students to browse current offerings, and find opportunities that align with individual interests and goals. Our web-based portal makes projects accessible and easy to find while also serving as a communications hub, sharing information about the model and connecting with key stakeholders including students, faculty, and partners. In addition to finding projects, students can apply for funding to support their project needs or opportunities to present at conferences or project showcases. These resources serve to incentivize engagement while helping students leverage their achievements in support of academic and professional goals.

While our Project Portal is designed to resonate with students, it is equally important to engage mentors. The work of incentivizing key stakeholders should not be minimized, especially for Project Challenges that begin with community and partner needs, rather than faculty research. Since faculty's mentorship often follows their own research, engagement or scholarship, it is helpful to identify mentors with related experience and relationships. Fortunately, there are growing communities of faculty, mentors and partners with interest and expertise related to sustainable development and climate action. Leveraging networks and relationships while incentivizing and supporting engagement are important areas for development. While approaches for engaging faculty and mentors will vary, it is critical to ensure that all stakeholders are benefiting from engagement while contributing to the success and sustainability of the model.

At our university, many faculty already engage undergraduate students in their research and recognize the inherent value of mentorship. Yet pedagogical support is appreciated along with help developing projects and integrating them within courses, research programs, or even grant proposals. Our model has proven useful in providing faculty with PEARL, which is detailed in the next chapter, as a valuable tool for designing internships, independent study courses, and enhancing minors or signature offerings. Since student projects often require related materials or access to technology or software, we provide project funding support, but engage faculty mentors in preparing funding proposals and stewarding awarded funds. In this way, we help faculty build capacity for engaging undergraduates while at the same time supporting students in building their own capacity for growth and achievement.

While project opportunities and incentives are both critical to the success of our model, it is ultimately the outputs that will drive continued engagement. In addition to project deliverables, Project Challenges generate an endless supply of stories, reflections, and digital media that can be shared with both internal and external audiences. As a front door for your Experiential Learning system, your Project Portal should be both welcoming and inspiring, inviting stakeholders to engage and contribute while showcasing the impacts for individual students, their partners, and the communities and ecosystems that stand to benefit.

Getting Started: Creating a Project Portal

To build our Project Portal, we conducted a number of design sessions with our core team, assessing the current landscape and identifying priorities for resources and support. The following questions and steps can be adapted to guide your own design work and further developed to meet specific needs and priorities.

1. Clarify Design Goals

- What are your core priorities for supporting Experiential Learning?
- What are the unique strengths and challenges that frame your work?
- Identify key stakeholders who need to be engaged and/or supported.
- How will you meet their expectations and engagement needs?

2. Conduct an Environmental Scan

- What Experiential Learning is currently offered within your organization and by whom?
- Examine patterns of access and engagement.

- What students/youth participate and who is left out, and why?
- How does engagement connect with key commitments or priorities?

3. Early Visioning Work

- What will be included in your Project Portal? Define minimal requirements.
- Who will mentor your projects? Think through implications and incentives.
- How will you drive students to your portal and incentivize their engagement?
- What other resources will you provide?
- How will you support innovation and project creation?

4. Creating a Web•Based Portal

- What resources and support are available to design and manage your portal?
- Who are the primary stakeholders you will need to attract and engage?
- What information, resources and incentives will be critical to include?
- How will you access or leverage necessary resources or support?

5. Outputs and Deliverables

- How will you showcase successful projects?
- Consider stories, testimonials, and digital media to be shared with external audiences.
- How will you make your portal both welcoming and inspiring?

Reference

Huber, M., Heath, C., Baxter, C, and Reed, A. (2021). Using digital badges to design a comprehensive model for high-impact Experiential Learning. In Y. Huang (Ed.), *Handbook on credential innovation for inclusive pathways to professions* (pp. 418–434). IGI Global.

6 PEARL Engagement Framework

Connecting students with meaningful project opportunities via a Project Portal is an important first step. But it is only the beginning. In Chapter 4 we emphasized the importance of facilitating students' movement through the various stages of Experiential Learning, helping them activate experiences and integrate them with academic and professional goals. In addition to supporting student growth, such facilitation is critical for assessment, tracking outcomes, and using data to strengthen and optimize the system. The formalization of a facilitation framework is especially important for leveraging Experiential Learning as a catalyst for both individual and collective impacts. Namely, connecting students with community partners from diverse regions and ecosystems calls for a consistent framework that is sufficiently flexible and dynamic. This need is best facilitated with a technology-supported engagement framework rather than relying on the skill of individual teachers, faculty or mentors.

We developed PEARL as a flexible framework that supports student engagement across diverse project types and methodologies while also providing functionalities related to research and assessment. While PEARL was created with these specific purposes in mind, it has emerged as an exciting design tool for project creation, innovation, and broader impacts explored in the final chapters. As we write this book, PEARL is being translated within diverse contexts and educational environments including K-12, open education, and vocational programs domestically and in West and East Africa. The diversity of applications is exciting and serves as an open invitation to innovate and find inspiration in the stories of our partners shared in Part III.

DOI: 10.4324/9781003489337-8

But first, we must introduce PEARL and how it is utilized to support projects and Project Challenges through our own web-based model.

PEARL is based on the simple yet powerful idea that experiences are constructed by the learner rather than simply acquired. Through a series of prompted activities, students Prepare, Engage and Add value, Reflect, and Leverage their experience toward broader impacts. While working on individual projects either in-person or remotely with their mentors and partners, students move through a universal set of online activities corresponding with three phases of PEARL. See Figure 6.1 for an outline of the phases and corresponding activities. Like our definition for Experiential Learning Network (ELN) projects, we keep PEARL as lean as possible, requiring a minimal number of activities to make the steps simple while offering the most critical facilitation and assessment. Each phase of PEARL is described below. Note that PEARL culminates in the earning of a digital badge. Badges are an exciting extension of PEARL and add dynamic functionalities. Our digital badges are detailed in Chapter 7. However, it should be noted that PEARL can be offered in conjunction with Project Challenges and community partners without the addition of badges.

P for Prepare

Preparation is about getting ready for an experience. For our Project Challenges we include the following Preparation activities which are detailed in the "Getting Started" section at the end of this chapter.

1. Setting goals and intentions for skill and competency development
2. Learning about the partner organization or the challenge
3. Exploring related SDGs
4. Any specialized skills or knowledge required for the project
5. Developing a plan for the project deliverable

Preparation work helps students build a foundation for engagement while supporting assessment related to key skills and competencies. It also prepares students to engage with their partner organization and establish a plan before getting started. As students work through Preparation prompts they are instructed to upload responses through forms and documents. As they complete their Preparation work, they support assessment. To be clear, Preparation can extend beyond our process. When professors integrate projects and PEARL within

Preparation Phase

Engagement Phase

Reflection Phase

Figure 6.1 ELN digital badge phases and activities as displayed on the ELN website.

courses and syllabi, we work with them to identify where Preparation ends and the next stage (E and A) begins. They are welcome to add additional Preparation activities as part of their course, but the PEARL sequence remains consistent across all projects and course structures.

E for Engage and A for Add Value

Once Preparation is complete, students move to the E and A phases which stand for Engage and Add value. We have combined these steps to ensure that the focus on Adding value is driving Engagement. This ordering is especially critical for Project Challenges. Although students are often drawn to a particular challenge because of the related technology and alignment with their major or professional interests, it is important to keep students focused on the importance of making a meaningful contribution through their project output. This is also critical for helping them persevere when things get difficult. Project Challenges are challenging by design. It is only when students work through challenges that they develop the skills and competencies that are so valuable to employers and the world. What is most important is that they honor their commitment to produce something meaningful, even if the outcome is significantly different from what they had anticipated.

Most of E and A (Engage and Add value) are happening under the direction of the project mentor, outside of PEARL activities or ELN oversight. The phase culminates in the uploading of curated evidence and some sort of approval by the mentor, and often the partner. ELN provides templates that are available for the students to use, but we encourage them to produce evidence that is meaningful for their project while also resonating with key audiences. In addition to posters and slide decks, students have uploaded videos, reports, business pitches, and even design sketches. Because our projects require mentorship and collaboration, students receive feedback throughout the process, and sign-off by mentors and partners to ensure adequate vetting. With that said, ELN staff review submitted evidence and provide formative assessment feedback, following a rubric that is provided within the workflow. It is important to note that the uploaded evidence gets attached to the digital badges that will be discussed in the next chapter. This provides an incentive for students to ensure that their final projects are appropriate and worthy of sharing.

R for Reflect and L for Leverage

Although ELN projects are technically complete with the submission of curated evidence, R and L (Reflection and Leveraging) are important steps in helping students get the most from their experience while supporting dynamic assessment. Through Reflection activities, students develop narratives that connect their experience with coursework and professional goals. The ability to effectively talk or write about their experience is something that students often find challenging and lack related practice, yet can be critical for securing valuable opportunities. As students practice talking or writing about their experience, they internalize the narratives and gain new insights and perspectives that are in turn related to 21st century skills. To support Reflection activities, we direct students to watch a video that shares a process for creating compelling stories about their experiences and achievements. It guides them in examining their experiences through the lens of key audiences including employers and graduate programs, then developing narratives around competencies and outcomes. Students report benefitting from the narrative development exercise and often identify this activity as being especially helpful.

The final phase of PEARL is Leveraging, which involves building on an experience toward greater impacts. While leveraging is important to getting the most from an experience, students (and the rest of us) are not particularly good at it. They have a tendency to quickly move onto the next activity without building on what they have created or accomplished. As a final step of PEARL and earning an ELN badge, students are directed to integrate their experience within their resume, and consider how they can deepen engagement or related activities in support of academic and professional goals. We also prompt them to consider how their partner could leverage their engagement toward broader impact. We have recently created a PEARL Award that provides financial support to help students leverage their ELN project or Project Challenge, going farther or deeper with their engagement in support of transformative educational or professional goals. Applications for our first round of PEARL Awards were impressive and included proposals for travel-based research and service, independent research, and participatory scholarship. This suggests exciting potential for supporting and incentivizing further engagement through Leveraging projects to catalyze even greater impacts.

Together, PEARL activities support engagement across diverse projects, mentors and partners, helping students get the most of their experiences while supporting academic and professional goals. PEARL also supports dynamic assessment, with students uploading evidence and data through forms and documents as they work through the various steps and activities. Because every student who engages with PEARL shares their unique student identifier, we can compare project completers (or badge earners) with groups of similar students who have not completed projects or engaged in formalized Experiential Learning. Using this assessment approach, we are able to examine the contribution of project completion to student persistence and retention, while also examining the importance of mentoring, community engagement, the Sustainable Development Goals and other components of our model, supporting experimentation and utilizing Experiential Learning as a targeted intervention.

In addition to dynamic assessment, PEARL has emerged as an exciting instructional tool, supporting the design of new or re-imagined courses. In Chapter 8, we discuss the benefits of curricular versatility and ways to integrate PEARL and Project Challenges within diverse course structures and formats. Other uses of PEARL are shared through the case studies presented in Part III, featuring our community partners who are translating the ELN framework within their unique contexts and educational environments. PEARL is supporting innovation challenges in a redesigned Basic Education curriculum in Enugu Nigeria, capacity building for youth in Nakivale Refugee Settlement in Uganda, and an emerging program for Open University students in Zimbabwe. As it is translated and adapted for new environments and audiences, we will discover new ways for supporting and assessing Experiential Learning while empowering youth to add value and leverage their engagement toward broader impacts.

Getting Started: PEARL

Now that readers have been introduced to PEARL and its functionalities related to facilitation, assessment, and project creation, the framework can be translated for specific organizational needs and context. Remember that at its core PEARL is about helping students get the most from their experience while also contributing to robust research and assessment. You can use the following outline as a foundation for translating PEARL and digital badges, which will be discussed in Chapter 7.

Preparation Phase: (P)repare

- Setting goals and intentions.

  We have students select two professional competencies from a provided list; we use the NACE Competencies but you can choose whatever is most relevant for your purposes. We prompt students to revisit the same list of competences in the Reflection Phase of PEARL but it is important for students to begin their project journey with intentionality and making an initial connection between the project and their academic and career goals.

- Exploring the SDGs and project partners.

  For our Project Challenges, we have students explore the SDGs and their assigned or selected partner. By making explicit connections between the work of the partner and related SDGs, we help students create a foundation for their subsequent engagement. Assignment prompts can include templates or instructions for students to upload an essay or file.

- Specialized skills and context.

  Since every project is unique, we encourage mentors to add any specific preparation activities that are helpful or necessary for setting the project and student up for success. Specialized preparation may be assigned through associated courses or discussed individually with students through consultations.

Engagement Phase: (E)ngage and (A)dd Value

- Typically, students spend the largest portion of their time in this phase, working on their projects and getting feedback and assistance from their mentors and partners. This phase culminates in the submission of their final project outcome as agreed upon during the planning phase.

- Expectations related to engagement, communication with partner, and nature of deliverable should all be clarified early in the process, with support from a mentor or facilitator, especially if they are completed outside of a course or formalized structure.
- When developing curated evidence for submission and integration with the digital badge, students should strive for resonance with key audiences such as employers, graduate schools, or other keepers of high-value opportunities.

Reflection Phase: (R)eflect and (L)everage

- In this final phase, students revisit the two competencies that they set goals around during their Preparation work and reflect on their growth via a pre/post assessment using the NACE Competencies.
- Next, we have students watch a video that guides them in constructing compelling narratives about their experiences and achievements, connecting with the priorities and needs of key audiences, emphasizing NACE Competencies and VALUE Rubrics (AAC&U).
- Based on the process shared in the video, we ask students to tell the story of their project experience, reflecting on their journey and connecting with stakeholders of interest. Student stories are valuable outputs for sharing with both internal and external stakeholders for further capacity building. We get student permission for sharing stories and other related data as part of the PEARL workflow.

7 Leveraging Digital Badges

With Project Challenges and PEARL, we can connect students with community partners doing the work of sustainability, and support meaningful contributions through project outputs. This alone is exciting. But there is one more critical component necessary for transforming Experiential Learning into a catalyst for broader impacts including climate action – connecting PEARL with digital badges.

What are digital badges? They are a form of alternate credentials that serve as proof of an acquired skill or competency. Think of a digital icon in the shape of a badge that showcases an achievement and can be shared with an employer, graduate school, or other valued audience. Much shorter than degree or certificate programs, micro-credentials and digital badges are awarded when learners demonstrate a threshold level of competence. They allow for customization and personalized learning as students work to develop skills and badges that are important for their unique academic or professional goals.

What is the difference between micro-credentials and digital badges? While this distinction may be defined differently in different spaces, on our campus, micro-credentials are credit-based and displayed on student transcripts while digital badges are co-curricular. For the ELN model, we utilize digital badges. Although they are not explicitly tied to courses, they can easily connect along with corresponding PEARL activities, allowing for seamless course integration. Credit can be awarded in addition to a badge, but the assignment of credit and grades are the responsibility of the instructor and the associated department. The decision to award badges rather than micro-credentials supports curricular versatility, detailed in the next chapter, while providing an

DOI: 10.4324/9781003489337-9

option for project completion outside of coursework, thus supporting our commitment to access and equity.

While digital badges and micro-credentials are being used to formalize achievement across a variety of domains and applications, we use digital badges to showcase project outputs and 21st century skills and competencies. When a student completes PEARL, they earn a badge, and their final project outcome gets linked or attached as clickable evidence. See Figure 7.1 for a listing of ELN's digital badges issued to those who complete Project Challenges. Students can post their digital badge icon on their digital resume, LinkedIn profile, e-portfolio, social media account, or any other digital place or space. The viewer will see the icon and can click on it. Upon clicking, details about the badge will be viewable along with a link to the project outcome submitted by the student and vetted by their mentor. Our badges are a great way for students to share a sample of their work and provide tangible evidence or proof of skills and competencies that they claim to possess.

Similar to the other dynamic components of our model – namely, Project Portal and PEARL – our digital badges offer both individual and collective benefits. For students, badges are a symbol of achievement and can be shared with key audiences. As they approach the completion of PEARL, students are encouraged to create a LinkedIn account and are directed to the Career Design Center for help. They are also instructed to add their badge to their résumé and utilize it for career development, leveraging reflection and their developed narrative. Students can show their badge to potential employers or graduate programs, but they can also talk about their experience and the specific skills and competencies they developed.

For young people in environments where experiences and jobs are particularly hard to come by, the potential for digital badges is even more exciting. The badge itself can be valuable, especially if associated with highly esteemed programs or designed with involvement from key stakeholders or employers. And if the attached evidence is especially compelling or powerful, that too can become a driver for opportunities. For example, if students can share evidence of coding work, or mastery of specific technologies that are in high demand, the badge functions as a portfolio and demonstrates their potential to add value. When we consider this potential through the lens of human capital or workforce development, we can view digital badges and micro-credentials as dynamic stepping stones, able to bridge gaps between

Figure 7.1 ELN's suite of digital badges given to those who complete Project Challenges.

formal education and emerging areas of need and innovation. Badges empower students to customize their degrees and coursework, demonstrating evidence of doing in addition to learning and knowing.

While our badges are useful to students who earn them, they have other functionalities that are equally exciting. By connecting badges with the PEARL framework, we can make assessment dynamic, activating potential for supporting and optimizing both individual and collective impacts and tracking progress toward identified goals. This potential for dynamic assessment is a new idea that warrants further explanation.

Let's begin with the assessment of student growth. As students work through PEARL activities we collect related data related to intention setting, pre and post assessment of competencies, reflection, and other metrics connected with our model design. We can examine differences between those who have completed PEARL and earned badges, and groups of similar students who have not engaged in Experiential Learning. This allows us to explore the impacts of the model in relation to key education metrics such as success and retention. We can also drill down to particular groups of students and demographics to understand differential benefits or ensure that target groups are sufficiently engaged while informing project creation and helping to optimize the efficacy of the model. Since students are encouraged to display their badges on LinkedIn, which is a public market for professional growth and employment-related networking, we can also use badges as trackers to learn about how students leverage them to support job success and mobility.

But rather than simply tracking our badges to see where they take our students, we can create or facilitate opportunities for those who earn them, including higher-level experiences or uber badges that allow students to "stack up" to bigger competencies or skills. We can connect badges with degree programs or specializations or invite in employers or partners with related needs or interests, creating dynamic pipelines around badges. We could support related research and innovation and bring in investors to offer incentivization and support. With these tools and resources, we can create dynamic ecosystems of opportunities with badges serving as connectors and young people empowered to build their own capacity while addressing our most pressing needs and challenges.

Getting Started: Digital Badges

While our utilization of digital badges is exciting, it is a dramatic departure from current uses that prioritize gamification and the (presumed) inherent value of the badges themselves. It is important to note that while badges are useful, they are not magical, and will not deliver transformational results simply by employing them. This runs counter to assumptions that are driving the embrace of micro-credentials and digital badges across diverse educational programs and offerings. Despite elevated expectations, badges do not automatically drive engagement, nor do they support assessment or the creation of value. Each of these functionalities and other priority needs must be developed and supported individually. The following design work can help guide your utilization of digital badges or micro-credentials and should be adapted to meet your individual needs and contexts.

Facilitating Engagement

- Consider connecting Project Challenges and PEARL activities with digital badges; how might you support students as they work through PEARL? Consider the most effective and efficient ways to administer badge activities.

Assessing Impact

- Consider comparing students engaged in PEARL activities with similar groups of students who haven't participated in Experiential Learning. How could you identify a comparative sample/s? Are there resources or expertise you might leverage? What assessment questions are most important to explore? Consider measures of student growth and success, as well as collective measures. Also consider the importance of stories and qualitative data as outputs of your system.

Creating Value

- As you think about attaching badges to projects and PEARL, consider how you might create interest around stakeholder engagement. Can you connect badges with some high-value opportunity as a way to incentivize student participation, or perhaps offer resources such as funding or recognition? Also consider inviting key stakeholders to help design and/or endorse the badges as a way to increase interest and engagement. Consider connecting your badges to other programs or offerings as a way to create pathways or pipelines. As you consider these ideas, check with students and other stakeholders to make sure they are sufficiently attractive since their value will lie in their ability to drive engagement. Lastly, choose the names/titles of badges carefully. Names matter and should connect with the focus of students' projects or project outcomes while also resonating with key audiences.

Practical Considerations

- Does your organization currently issue any type of micro-credential or digital badge? If yes, who is issuing them and for what? How can you build off of what exists? If necessary, explore available badging and micro-credentialing platforms. There are a growing number of both paid and open-access options. Note functionalities that will allow you to customize and brand your badges along with differences in supporting dynamic assessment through meta-data. Note that while badging platforms and companies are important, the badge is only the shell. You will need to build out the facilitation activities that will support the awarding and earning of the badges. Ultimately, you must decide how you will support students as they work through PEARL and meet expectations for earning a badge.

8 Curricular Versatility

With our Project Portal, PEARL, and digital badges in place, we can invite the engagement of faculty, students and partners. But to fully activate the model, we need to integrate these components within courses and programs. The decision to make our badges co-curricular – not directly attached to course credit or student transcripts – offers exciting possibilities. Because PEARL modules are flexible and relevant across diverse project types and methodologies, they can fit within virtually any course or programmatic structure, allowing for seamless integration while leveraging the model's ability to support and assess both individual and collective impacts.

The notion of curricular versatility is exciting for education and warrants further explanation. Because our model is so flexible, it fits within almost any curricular structure or focus. And because our badges are not credit-based, we can leave it to faculty and departments to add requirements and vet related expectations. We handle the badges and the PEARL activities, while faculty and departments do the rest. This arrangement allows for maximal flexibility while providing consistent facilitation and supporting robust data collection and assessment. And because almost anything can qualify as a project, as long as it is mentored, collaborative, and results in something inherently tangible and meaningful (our core requirements for inclusion in our Portal), we can work with faculty and departments and external partners to customize projects to fit their needs and priorities.

This arrangement allows for exciting collaboration and utilization of our infrastructure to support strategic goals and priorities for both the university and individual schools and departments. Because PEARL modules are supported by ELN team members through

DOI: 10.4324/9781003489337-10

holistic mentoring and formative feedback, faculty can focus on providing students with course-related instruction and engagement. Students, in turn, enjoy the resources and benefits that ELN offers, including project funding, digital badges and opportunities to share their project outputs with key audiences through poster sessions and conference presentations.

When we designed the model, we envisioned integrating projects, PEARL, and badges within diverse types of courses including independent study, research, study abroad, internship, service learning, or virtually any course that could accommodate a project or Project Challenge. We needed the model to connect with existing projects while also supporting the design and development of new offerings. Using PEARL to identify activities and content associated with each stage of engagement (Prepare; Engage and Add value; Reflect; and Leverage), with a focus on the project outcome, we could work with departments to provide students with exciting opportunities to support their learning while developing evidence of skills and competencies through digital badges.

While the ELN model can be integrated within any course structure or focus, our SDG Project Challenges require additional project features including a focus on the SDGs and engagement with community partners doing the work of sustainability and climate action. In developing this category of projects, we were hopeful that there would be interest among UB faculty but also colleagues from our State University of New York (SUNY) System. In 2024,we received a SUNY Innovative Instructional Technology Grant (IITG) to scale our Project Challenges across our 64-campus system, creating opportunities to invite engagement of new faculty and their community partners. In addition to providing funding support for our SUNY partners, the grant created an opportunity to enhance our model through the utilization of ArcGIS StoryMaps as a tool for Reflection and Leveraging along with supporting dynamic assessment.

ArcGIS StoryMaps is a story authoring web-based application that allows the designer to share maps in the context of narrative text and other multimedia content. By giving students access to StoryMaps and other ArcGIS mapping tools, we hoped to support student reflection and integration of exploration activities while expanding impacts for our partners. In addition to their multimedia functionalities, StoryMaps allow for the development and sharing of compelling narratives and can accommodate co-created maps featuring in-situ data provided by partners. The relevance of mapping (with) our partners cannot be overstated and is discussed in both Part III and in the Epilogue. Although we have just begun experimenting with StoryMaps and ArcGIS data gathering tools, we are excited to share these functionalities with our SUNY partners and ELN students. The following are courses that are currently being supported through the SUNY IITG Grant and are integrating Project Challenges and the ELN model including ArcGIS StoryMaps and digital badges.

Rwanda Project Challenge: Buffalo State University

Buffalo State/Anne Frank Project has been taking students to Rwanda for over ten years. This highly immersive experience includes a deep examination of Rwanda's history with specific emphasis on the *1994 Genocide Against the Tutsi in Rwanda* (pre-1994, during the Genocide, reconciliation process). Once the students have experienced a respectful foundation of Rwanda's culture, they help to facilitate the training of Rwanda school teachers in AFP's story-based learning model using stories as the curricular platform in their classrooms. AFP trains over 100 teachers each summer from over 40 different schools. Working with devoted partnerships created over the past decade, the students are immersed in unique, challenging, and transformative experiences. Students return to the Buffalo State campus to share their Rwanda stories at Buffalo State's annual Social Justice Festival and through building short plays that tour local Buffalo Schools. This curricular cycle provides a story-based platform for students to prepare appropriately for their international experience, engage directly with Rwandan people, their culture, their institutions, fully process their unique east African experiences, and share what they've learned with the local community: Literally bringing their transformative Rwanda experiences home to Buffalo, NY. Following the SDG's of Quality Education (4) and Peace, Justice, and Strong Institutions (16), our student team will work under faculty guidance to study and contribute to this unique partnership between Buffalo State University and Rwanda, Africa. This program aligns seamlessly with the ELN SDG Project Challenge and Digital Badge model. Students will complete PEARL modules as they work through the program and will develop StoryMaps to showcase partners and build further capacity for the program. The ELN Model will allow for robust assessment, allowing the AFP to explore and demonstrate impacts of engagement with regard to student success, retention, growth in NACE competencies, and other key indicators.

Honduras Project Challenge: SUNY Oswego

SUNY-Oswego's Honduras (Bay Islands) Marine Wildlife Ecology education abroad program has brought students to the Roatan Institute of Marine Sciences (RIMS), Honduras, for over 15 years. The experiential learning program combines academic instruction, hands-on fieldwork, self-reflections on questions of identity and values, and intercultural learning to better understand the complexities of conservation and animal behavior research as they relate to their entanglement with socioeconomic and human cultures.

Coral reefs worldwide are declining in health due mainly to overharvesting, pollution (e.g., nutrient runoff from fertilizers and detergents), climate change (elevated water temperatures and increased storm intensity), and disease (linked to stress from pollution and climate change). Students develop the skills and competencies supporting the SDG Life Below Water goal to conserve and sustainably use the world's ocean, seas, and marine resources. Our collaboration with colleagues at the Roatan Institute of Marine Sciences partnering with the Roatan Marine Park provides students with insights and experience working with international education and conservation programs on the largest Bay Island in the Western Caribbean.

Education abroad preparation work for participating SUNY Oswego students starts two months before traveling abroad. It includes completing the PEARL Preparation steps and online instruction to prepare students for active participation in field research in Honduras. Online instruction also introduces the strategy of capturing public support for marine conservation by leveraging charismatic species, such as coastal bottlenose dolphins, to highlight the importance of protecting marine habitats. A series of scaffolded online lesson modules introduce students to the life history characteristics and identification of the indicator species of fishes, turtles, coral, algae, and invertebrates forming the ecological communities surrounding the island. Interactive lesson modules also introduce students to the bottlenose dolphin (*Tursiops truncatus*) pod members they will study and work with while in Honduras. Pre-departure activities also introduce and guide reflection journaling to prepare students for self-reflections throughout their journey.

After completing the two-month online instruction, students travel to Roatan, Honduras, for the program. This program has required advanced planning and risk management preparation beyond typical SUNY Oswego education abroad experiences. The program location, which itself is safe but is in a country known for risk of violent crime, places students directly in the water. This has required a low faculty to student ratio so as to ensure water safety, as well as certain pre-travel precautions for students in order to participate safely. The SUNY Oswego instructors' expertise, experience level, and on-the-ground connections with RIMS since the year 2000 are essential to make this program work institutionally.

Upon arrival students take part in a study analyzing complex social networks among dolphins. The nested nature of bottlenose dolphin alliances is vital to facilitating their foraging strategies, managing competition, and avoiding predation in complex coastal environments. Working with the RIMS pod allows students to better understand the complex behavior of social networks using a combination of network analysis and experiments involving synchronous innovative behavior trials. Additionally, students will explore and discuss conservation issues that affect the future of Roatan's coastal bottlenose dolphins.

Completing the online indicator fish ID activities prepares students to conduct roving diver/snorkeler reef fish indicator species surveys at different sites on the reef surrounding the island of Roatan. Analysis of survey data compares reef health based on geomorphology and benthic conditions (i.e., algal cover versus coral cover) of patch reefs and fringing barrier reefs and their proximity to different levels of environmental disturbance.

RIMS, in partnership with the Roatan Marine Park, the Bay Islands Conservation Association (BICA), and the National Institute of Forest Conservation and Development (ICF) Honduras, established the initial coral nursery program in the Bay Islands. The coral tree nursery project at RIMS aims to help restore local populations of the once-dominant staghorn coral (*Acropora cervicornis*) and elkhorn coral (*Acropora palmata*). Disease, hurricanes, and climate change reduced the abundance of these species throughout the Caribbean by nearly 90% over the past 20 years. These branching corals are foundational to reef structure and provide critical habitat for fish and invertebrates. Students will assist in various aspects of the nursery program, including maintenance, fragmenting new genotypes to expand the genetic diversity of the nursery, and outplanting. SCUBA-certified students will help clean coral fragments on the nursery frames suspended in 40 feet of water and assist with outplanting on the reef. All students will assist with microfragmenting coral to create propagules for the nursery and complete an outplanting exercise that simulates outplanting on the reef.

Dependent upon the focus of their ArcGIS digital story map, students will earn either an Ecorestoration or Mentored Research badge. In addition to sharing their badge on the social media site of their choice (e.g., LinkedIn, ePortfolio), they will be encouraged to present their work at SUNY-Oswego's annual Quest Symposium.

Designing for Pollinators: Enhancing Biodiversity and Multispecies Community in the Built Environment – University at Buffalo

Students in the sophomore architectural design studios are developing design proposals for a community center aligned with the Brooklyn Botanic Garden (BBG). As part of the semester's project, they are designing and fabricating pollinator habitats that will be included in a public exhibition at BBG tentatively titled "Natural Attraction: A Plant Pollinator Love Story," slated to open in May 2024. The project engages most clearly with the UN SDG 15: Life on Land, as well as 11: Sustainable Cities and Communities, 13: Climate Action, and 4: Quality Education. As part of this design process, students will need to visit the project site in Brooklyn to make and document observations through mappings, recordings, and drawings that will be important for them to analyze the site in order to develop their design proposals. During the visit to NYC, we also plan to engage the BBG's Director of Interpretation and Exhibitions, and Vice President of Education and Interpretation, as well as other NYC-based experts in ecology and art in a "review" – that is, a formal conversation/discussion with experts for students to receive feedback/critique on design ideas (which is a central mode of learning for students of architecture). The site visit and NYC review will require three days, including travel time.

Getting Started: Curricular Versatility

Figure 8.1 outlines the most common way the ELN model has been integrated into existing courses and programs on our campus.

Given that there are many options for customization, below is a list of questions and ideas to consider for integrating projects, PEARL and digital badges within courses and programs of study.

Mentored Projects

Which courses or programs already contain projects as a key component? Consider starting with existing offerings as a foundation for further development. Identify faculty or mentors who are open to creating new projects and introduce PEARL as a dynamic tool. What are the existing Experiential Learning opportunities that could be translated into a project (mentored, collaborative and result in a meaningful project outcome) with a few minor modifications? Consider study abroad and internship courses which are often particularly well suited and represent a great place to start.

Structure, Details and Design

When designing integration, consider whether students should complete a project or badge as a key part of the course or program. Consider the implication of making it required or including it as an alternative to an exam, traditional paper or assignment, or as extra credit. Does the course or program allow for an optional project component to be added to the existing content for students who want to go above and beyond/do more?

Accountability

How will you hold students accountable for completing PEARL and perhaps earning a badge? We find that grades are particularly powerful in holding students accountable and encouraging them to complete their project and badge, but there are certainly other ways to accomplish this.

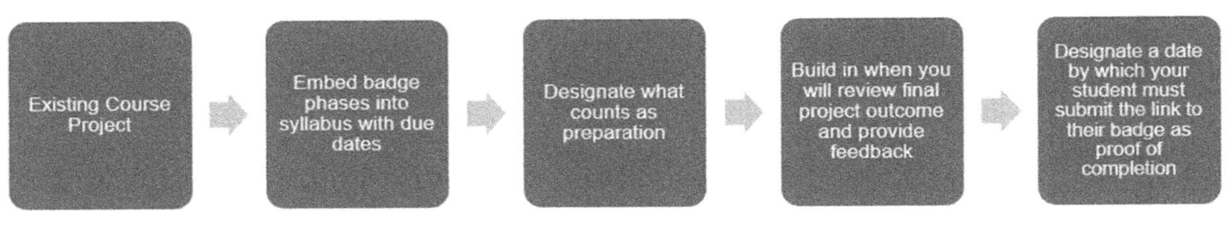

Figure 8.1 A visualization of how the ELN model can be integrated into a course with an existing course project.

Fit

What makes the most sense for your individual course or program? The ELN model is intentionally flexible to allow for customization and experimentation. Don't be afraid to jump in and play with curricular integration, trying out different ways to connect class activities with PEARL, or constructing a new course around the framework. Then, reflect on what went well, what did not work, and make changes for the next iteration.

Part III

Partners, Projects, and Students

9 Empowering Women Farmers in Uganda

Written with Deo Mbabazi, Executive Director of Biringo Women's Development Association (BIWODA)

Farming is the foundation of women's livelihood in Biringo Uganda. In many ways the region is well suited for this practice. The soil is fertile and the conditions support diverse crops including cereal, legumes, and root varieties – especially Irish and sweet potatoes and some fruit trees like avocado and pawpaw. Agriculture is the lifeblood of women in Biringo and other rural parts of Uganda, especially widows who have no other means of providing for themselves and their families. In Biringo, women farmers survive on approximately 2,000 Ugandan shillings a day, which is less than a dollar. Without credit or collateral, it is impossible for them to borrow money or start their own businesses. They remain in a perpetual state of vulnerability, unable to support basic needs including healthcare or education for their children.

Deo Mbabazi was raised by women farmers in Biringo. He refers to them collectively as his Mothers. Deo's actual mother gave birth to 12 children but only Deo and five siblings survived. His father left the family when he was still young and his mother struggled to support their basic needs including food and shelter. When Deo completed his primary level schooling, there was no money to attend secondary, the equivalent of high school. Although secondary school is technically free and compulsory in Uganda and throughout much of Africa, there are costs including those for uniforms and books that make it inaccessible for many poor families like Deo's. But Deo's mother was part of an informal women's group made up of farmers from the village. She talked to the women about Deo's schooling and they decided to pay for his fees. Deo's mother eventually paid them back by weaving and selling mats made from papyrus reeds, but the

DOI: 10.4324/9781003489337-12

women's group continued to provide support until he finished secondary school.

Deo recalls a time when his mothers came to visit him. He remembers clearly that it was a Tuesday when they made the long journey to the school where he was boarding. Although they had virtually no money between them, they left Deo with 200 shillings, enough to buy a pancake or a cup of porridge. They also brought him a handkerchief – it was white with red stripes. Deo kept this day in his mind since their visit meant so much to him. As he continued his studies and started developing skills, he decided to give back to his community and applied to the university to study education, eventually earning a degree. After graduating, he returned to the village, to the women's group that had supported his schooling. He set out to put it in order, looking for ways to develop and sustain its operations. He held meetings and worked to formalize and rename the center, now known as Biringo Women's Development Association (BIWODA). Deo envisions a community of successful women farmers with skills that allow them to live empowered and healthy lives.

The need for Deo's vision is compelling. While the land in Biringo has long supported farming, the work is getting harder to sustain. Due to overuse, hilliness, and fragmentation, land is becoming less fertile and many native crops that are subsidized by the government are no longer surviving in the harsh conditions. And with continued cutting of trees for cooking fuel, deforestation is becoming increasingly problematic. As land is degraded, extreme weather crises are more frequent and include prolonged droughts and destructive floods, making it harder to cultivate crops and support the needs of families and children. When Deo considers the root cause of women's vulnerability in Biringo, he focuses on poverty, SDG 1. He explains that women cut down trees for firewood because there are no alternatives – they need to provide food, and trees are all that they have for fuel. He reminds us that in his country, particularly in rural communities like Biringo, women do not have the right to own property or land. They are considered property themselves. So a property cannot own a property. This is the mentality in his village and it makes life particularly difficult for girls. Because they are expected to do work at home, and there are endless chores, school attendance is chronically low and girls eventually drop out without progressing to advanced levels where they can learn skills and access opportunities.

According to Deo, sustainable agriculture is the solution along with training and education. Through BIWODA, Deo is working to provide both. He is committed to ensuring that each woman farmer has a goat. The droppings can be used as fertilizer to grow vegetables and surplus crops can be sold for income. And by planting vegetables, the women can create more ground cover, reversing patterns of deforestation and degradation of soil. But Deo wants to go even further. Through training and the development of educational resources, he wants to introduce the women of Biringo to local technologies that can mitigate the effects of climate change while generating revenue and empowerment. Deo is putting his vision in motion. He has met with local leaders and secured two acres of public land for a training and education center, along with a future primary and secondary school. Through the new center he will introduce women to community technologies, focusing on those that address priority needs and challenges. Although Biringo women lack education and skills, Deo believes that with mentorship and access to technology, they will become empowered and contributing members of their communities.

Deo sees college students as instrumental to achieving his vision. He has been engaging with students from the University at Buffalo (UB), inviting their collaboration around issues of poverty, sustainable agriculture, and cooking fuel. Deo views these areas of focus as fundamentally interrelated and has framed his Project Challenge to include the following: (1) engaging students in discovery, needs assessment, and documentation of land conditions toward piloting new farming techniques with a focus on revenue generation and supporting community needs; (2) exploring models of social entrepreneurship and collectives that have been successfully implemented in rural communities of women, and develop related training and resources aligned with identified opportunities; and (3) working to ensure access to affordable and reliable modern energy sources.

Although Deo emphasizes the importance of all three project options and their interconnectedness, he gets most excited about the potential of sustainable cooking fuel. He explains that Biringo women currently use firewood to support their cooking needs. Although firewood is available in the villages, it is very expensive with logs costing 10,000 shillings each. Since most women cannot afford this expense they end up traveling by foot into the hills to cut down trees. This makes them vulnerable to sexual harassment and even rape. Alternative energy sources include eco-stoves which are more efficient and use less fuel allowing women to save money and time. They also offer environmental benefits and can use renewable fuels such as briquettes. BIWODA women are eager to explore alternative energy sources and develop plans to support a shift in utilization.

While Deo's Project Challenge is compelling, it requires students to resonate with the opportunity and activate its potential impacts. In the ELN, we have a saying that rather than helping students find their dream project, we help projects find their dream students. In this case, Deo's Project Challenge worked its magic as soon as it was posted, attracting the attention of Ashley Brito, a junior Environmental Engineering student with a compelling story of her own. According to Ashley, Deo's vision immediately captured her attention and evoked a deep resonance with her own childhood in the Dominican Republic. Living in a small pueblo, Ashley and her family faced significant challenges securing a consistent and reliable energy source. Having spent her formative years in the Dominican Republic before immigrating to the U.S. at the age of six, she understands the hardships associated with the lack of essential resources for a comfortable life.

For Ashley, Deo's project holds a personal significance aligning with the struggles she witnessed and experienced firsthand. Through her Project Challenge, she is driven by the aspiration to bring about positive change for the women in Biringo, aiming to alleviate the challenges posed by a scarcity of energy sources that constrain their quality of life. While Ashley is still completing her project, she feels that the understanding she's already gained through engaging with Deo and his work is fueling her desire to address the pressing needs of communities like the one she grew up in. Ashley's ultimate goal is to leverage the knowledge cultivated through this project and channel it back to her motherland. She envisions developing sustainable systems that will provide comfort and improve the quality of life in low-resource communities. Deo serves as a source of inspiration, and Ashley is motivated to follow in his footsteps. Through dedication and application of the skills she is acquiring, she hopes to contribute to transformative changes for communities facing energy disparities.

In activating Deo's Project Challenge, Ashley has been immersing herself in innovations related to biofuel, plant-based cooking methods, and renewable energy. Through early exploration of breakthrough technologies, and her own experience as an Environmental Studies student, she suggested that biofuel might be a good option for Deo to consider. Through dialog and reflection, Deo recalled innovations that were being implemented in a nearby city, transforming animal and plant waste into cooking briquettes. He was particularly excited about the use of papyrus reeds, a plant that is plentiful in Biringo, growing wild throughout the wetlands and quickly regenerating for a continuous supply. He shared the work of Simon Go Green, a sustainability innovator who is transforming a multitude of plant and animal-based waste. Currently Ashley and Deo are working on an innovation of carbonizing the papyrus leaves and chopped reeds without applying heat or any machinery by using the anaerobic digestion process. Once the result comes out positive, it can be used to make the mold for the briquettes. It will translate the technology in a sustainable and easily accessible manner.

Simon describes himself as a clean energy innovator and is currently producing high-quality charcoal briquettes from biomass to substitute cutting down trees in Uganda for charcoal production. He also provides energy-saving stoves for commercial and household use of biomass briquettes which are efficient and relatively cheap compared to other sources of energy in Uganda, and installs biogas systems for institutions and households. Simon is Director of Ecoverse Innovations Ltd., formally registered as a clean energy production and green energy services company. He shares that at Ecoverse Innovations Ltd. they also do skills training and technology transfer in communities and countries in Africa in clean energy and green services.

The possibility of skills training gets Deo especially excited. If the women farmers of Biringo could learn to make cooking briquettes from papyrus reeds and other biomass, they could have a safe means of cooking while also developing a skill that could generate revenue and support their priority needs. Deo has shared this with Ashley who is working to piece the model together. She is engaging with digital artifacts and information given by Deo and Simon, and integrating her own research on innovation within this space. Ashley's goal is to translate what she finds for the benefit of Deo and the women farmers of Biringo, but she hopes that her project will find additional audiences. Although she is still in the engagement phase, Ashley hopes to produce several outputs including a video with Simon demonstrating the process of briquette making, along with a detailed guide that explains the process and the necessary equipment and materials. The guide will include itemized costs and sufficient detail to support grant writing or fundraising. If time permits, Ashley hopes to identify foundations that might be interested in supporting a pilot initiative in Biringo.

When asked about her Project Challenge, Ashley shares that it is different from what she had expected. While her engineering education, particularly in the realm of energy, often focuses on resolving community issues through advanced technologies like solar panels and wind turbines, she recognizes the need for a distinct approach in communities like Biringo. Solutions must prioritize easily accessible and low cost materials – an aspect not typically considered in urban settings where ample resources and financial means are more prevalent. The process of researching technology innovations and translating models for local communities is something that Ashley now values and recognizes as important for sustainability. The more she works on this project the more hopeful she is about rural villages like Biringo but also more cognizant of the importance of communication, the value of persistence and problem solving, which are critical when the work becomes difficult or unclear. Ashley admits that the translation of her findings has been prolonged due to research challenges, particularly in identifying suitable materials for briquette production. However, she finds hope in the efforts of people like Simon and Deo, who are working in rural areas to apply innovative solutions to communities in need, which fuels her determination to find the best approach for the women of Biringo.

Before graduating, Ashley wants to get more students involved, engaging teams to work with community leaders to translate technologies and support implementation. She thinks the project will especially resonate with Engineering students from international backgrounds and those who want to make a positive change in the world. She points out that many young people want to make a difference and are actually yearning to move beyond coursework and studying. Working with Deo and Biringo has been a highpoint and it makes her eager to go even further as she thinks about the next stage of her education and career.

As for Deo, he continues to dream of better things for the women of Biringo. And he is excited about the possibilities for Experiential Learning, but not just for UB students, also for youth in Uganda who are eager to develop skills and competencies and learn new technologies to support their villages and families. He reminds us of the land that has been identified to house his new school and vocational center. While he is appreciative of the work that Ashley and other UB students have offered, he imagines what will be possible when local women and youth take on their own Project Challenges and badges. He assures us that he can be patient, that new ideas take time to take root and grow, especially in rural villages like Biringo. But he is confident that in time we will see the fruits of our labor and the potential for stronger and healthier communities and empowered women. Ashley shares this vision and plans to focus on sustainable energy when she graduates. She hopes to continue to work with Deo but knows there are many more organizations and communities in need of local solutions, including those back home.

10 Addressing Vulnerability with Community Technologies in Tanzania

Written with Kennedy Mahili, Founding Executive Director of Community Life Amelioration Organization (CLAO)

When it comes to the Sustainable Development Goals, Poverty is SDG 1. Although overall rates of poverty have decreased dramatically around the world, extreme poverty remains concentrated within the most rural regions and vulnerable populations. Viewed as one of the most critical SDGs, Poverty is also among the most challenging. Since the very poor tend to live outside of formal systems including education, healthcare, and the workforce, they are largely invisible. Creating effective solutions requires getting close to marginalized communities, understanding their needs and challenges and identifying opportunities for empowerment. This is the work of our partner Kennedy Mahili and his community-based organization, CLAO (the Community Life Amelioration Organization).

CLAO is headquartered in Mwanza, Tanzania, on the shores of Lake Victoria, near the borders of Kenya and Uganda. Through CLAOs programs and resources, Kennedy cares for the most vulnerable, mostly women and children living in nearby villages. Kennedy's work is highly personal. As a child he was among Tanzania's most vulnerable. And although he is no longer defined by his story, he recognizes the importance of sharing it with those who are willing to listen. Kennedy was born in Kemakorere village, Tarime District, in Mara Tanzania. His mother gave birth to 13 children but only 7 survived. Kennedy's family was very poor and illiterate. He was the only one to attend school. His father was an alcoholic and a very violent man. Due to this situation Kennedy's mother took care of him and his siblings and their life was very miserable. To pay for Kennedy's school fees, his mother made and sold local alcohol, which is illegal in his country. She was eventually caught and sent to

DOI: 10.4324/9781003489337-13

jail where she stayed for three months. This put an end to Kennedy's schooling and made life especially difficult. So he decided to move to Mwanza City in search of a job. Kennedy was only 12 years old at the time and lived on the street with only a piece of box to cover him at night. He had no money and spent his days visiting cafes searching for leftover food.

Kennedy looked to be a patient and kind young man. And one day a woman took him from the street to work as a house boy. She paid for his food and gave him a place to sleep. He learned many skills and due to his hard work was taken to a cafeteria to wash utensils and help the cooks. Kennedy earned 15 dollars a month and kept the money until he had enough to pay for school fees. When he was 15 years old Kennedy joined the ordinary level and then things changed again. He had used all his money and needed to find more work in order to pay for continued schooling. He was hired as a cook in a hotel based on his experience in catering service. He worked through the day and attended adult education classes in the evening and during the weekends. Kennedy was eventually able to complete his studies and was selected to join university where he earned a Bachelor's of Art in English Education. During university he had a Good Samaritan who sponsored his studies. He now has a Master's degree in Community Development, a farm, and a community organization of his own. Kennedy is a self-described Man of Feminism and a Fighter for Disabilities Rights. He is committed to making lives easier for the most vulnerable and to champion what is right.

Kennedy explains that in many African countries, especially Tanzania, there are three enemies or obstacles to community development: poverty, illiteracy, and lack of education. Since so many children drop out of school and there is no vocational training to prepare them for self-dependency, many live miserable lives without access to resources or opportunities. The plight of out of school youth in Tanzania and across East Africa is well documented along with the particular challenges facing girls. Although education is technically compulsory through secondary level, there are numerous obstacles that limit attendance and completion. The cost of books and uniforms are a common barrier, and for girls, household duties take precedence over school. Fetching water and firewood, caring for siblings, and other daily chores contribute to high dropout rates. Without education there are few opportunities to develop skills or get access to jobs or money to support families. Not surprisingly, out of school youth experience high levels of criminal behavior, addiction and mental health problems, destabilizing already fragile communities.

Along with out of school youth, Kennedy supports people with albinism. Tanzania has the highest rate of albinism in the world with 1 in every 400 people having the genetic condition compared to 1 in 15,000 across Sub-Saharan Africa. People with albinism have unusually light skin and extreme sensitivity to sunlight with prolonged exposure leading to sun damage, cancer, and blindness. The rate of albinism around the Lake Region, where Kennedy lives, is especially high, although accurate numbers are difficult to come by. Traditional beliefs hold that people with albinism are born with a curse and their body parts are used by practitioners of witchcraft and feature in rituals and spiritual practices. With few safe options, parents of infants with albinism often leave them at the houses of village leaders who take them to infant care centers, placing them with NGOs and other community organizations. Caring for these children is especially challenging because of their special vulnerabilities and needs. They are under constant threat from the sun and require protective clothing, hats and sunglasses which few can afford. Living in extreme poverty they have no way to provide for their families and few attend school due to constant danger from trafficking and violence.

Kennedy explains that out of school youth, people with albinism, and other vulnerable populations live primarily in rural villages and rely heavily on agriculture to support their needs. They are entirely dependent on the land and especially vulnerable to the effects of climate change. For people with albinism in Tanzania, prolonged drought in rural areas is forcing them to walk long distances to get water, making them vulnerable to victimization. The extreme temperatures are also exposing them to the sun and the likelihood of developing cancer. The situation facing girls is particularly dangerous. As rivers dry up and deforestation forces them to walk further to find water and firewood, they are at increasing risk of violence. And if girls are raped and get pregnant, they are unable to finish or return to school. As a result of climate change and destructive environmental practices, climate crises are continuing to increase, disproportionately affecting vulnerable populations.

Although policies and laws are enacted to protect vulnerable populations, they are often not enforced in rural communities. In places and ecosystems that are themselves vulnerable, there is competition for limited resources. It is up to community-based organizations like

CLAO to provide care but more importantly to find ways to empower vulnerable communities to generate resources to support their own needs and priorities. If marginalized groups can decrease financial dependency, they can negotiate their own rights while contributing to the sustainability of their communities and changing perceptions over time.

While clearly not easy, Kennedy insists that the possibility of self-sufficiency is entirely realistic. There are community technologies and small-scale entrepreneurship models that are well suited for youth and marginalized communities, technologies that can produce goods and services for which there are markets and sources of revenue and can at the same time mitigate impacts of climate change. Although the word technology is most often associated with computers and sophisticated equipment, technologies solve problems through some targeted solution. Kennedy trains vulnerable groups on using community technologies to address needs and challenges within rural villages where the most vulnerable tend to live.

If you ask Kennedy what the most transformative community technology is, he will answer brick making and show you a hand-powered press that he uses with youth and vulnerable populations. By filling the trough with local soil and a little cement, and putting pressure on the handle, the press produces high quality interlocking bricks that can be sold individually or used for construction projects including eco-latrines and water catchment tanks. Kennedy explains that sanitation is a big challenge in rural Tanzania. There are no septic systems and no pipes or public toilets. While governments now require families to have a toilet, many fail to comply and those who do often opt for pit latrines which are dug holes, some with cement around them. Kennedy explains that pit latrines need to be pumped every two years. The financial cost is significant but the cost to the environment is even greater. Frequent floods cause run-off and human waste gets washed into drinking water, causing widespread disease and sickness. And with few clinics and resources to pay for care, health problems can quickly escalate.

Kennedy also points out the relationship between sanitation and girls' education and maternal health. Because of the lack of suitable toilets and water and soap in rural communities, especially around schools, girls often stay home during menstruation and use unsafe materials to manage their periods. They use whatever is available including rags or bed stuffing, which is unclean and can cause infections. And with chronic absences they often drop out and are traded into marriage for a dowry for the family. Dowries are still practiced throughout rural Tanzania and include cows that are highly valued and valuable. Because unmarried girls are an expense to families that must provide food and shelter, marriage represents an economic opportunity and is the bedrock of community life. Girls can be married as young as 12 or 13 and often as the second or third wife, bearing many children with few opportunities of their own.

Kennedy explains that water and sanitation are part of a complex set of conditions that perpetuate vulnerability and poverty. Kennedy works with marginalized groups to construct eco-latrines from handmade bricks. Eco-latrines use dry composting to transform human waste into fertilizer that can be used for agriculture purposes. Eco-latrines do not use water, which is scarce in rural communities, and the waste is trapped in the brick enclosures with cement, so they cannot be compromised by rainfall or flooding. Although eco-latrines are a little more expensive on the front end, they do not require pumping which requires a separate fee. Kennedy is working to make them more popular and by engaging local groups of youth in constructing bricks and structures, he is building capacity for this technology which can be shared with other vulnerable communities.

Kennedy also uses bricks to make water catchment tanks that are being used in area schools. While Kennedy is eager to discuss the design and construction of his tanks, he is quick to share a beautifully rendered technical brief created by Hannah Ruth, a recent graduate of the UB Architecture program and a participant in the 2023 summer study abroad trip to Tanzania that featured engagement with Kennedy and CLAO. In addition to technical drawings, Hannah produced a graphic vignette storyline with photos from their trip and details of the water tank construction that the group participated in under the mentorship of UB Professor Chris Romano and SUNY Empire State Professor Dan Nyaronga.

Hannah's vignette explains that despite being adjacent to the generous fresh water of Lake Victoria, the rolling hills of nearby towns lack the infrastructure to pump water or power up the steep slopes, leaving thousands of residents without access to basic necessities. During the rainy season, water is directed from the roof into a suspended pipe that feeds into the tank, holding approximately 2,468 gallons of non-potable water for handwashing, toilets, farming, and other basic necessities for the local students and their families.

In response to the minimal availability of concrete in rural Mwanza, CLAO uses hand-operated compression technology to transform the abundance of rich local clay into a structural building element.

Through a year-long project, collaborating with colleagues from Geology, Environmental Studies, and companies dealing with cement and mineral composites, Professor Romano engaged his students in extensive exploration of the composition and structure of the blocks that Kennedy uses for his tanks, eco-latrines, and other infrastructure. UB students have been able to use a similar hand press, allowing for the development of early knowledge and experiences that lay the foundation for the study abroad trip and Hannah's continued engagement as she prepares for graduate work.

Hannah's architectural renderings and technical briefs are supporting a return trip planned for this summer. Through her continued work with Professor Romano and engagement with Kennedy, she is working to open more doors and get more people involved in these collaborative efforts. Among her specific goals, she hopes to examine how to build a more durable tank using improved stabilized soil mixtures and structural principles; how to size the water tank based on both local rainfall totals as well as occupant demand; and how to graphically communicate low-cost, locally sourced, locally employed water tank construction to those without technical training. Hannah shares that she hopes to use architecture as a lens through which to connect the world and bring forward the voices of its people. She plans to branch across a variety of disciplines and use her technical skills to develop universal visual language through which we can address collective global priorities as described in the UN Sustainable Development Goals.

Meanwhile, Kennedy is busy dreaming up his own Center for Experiential Learning and Community Technology and is eager to host more study abroad classes and welcome community groups from throughout Tanzania and beyond. While he will happily train anyone interested in utilizing community technologies, he remains committed to serving the most vulnerable, and hopes that his approaches can end the cycle of poverty and vulnerability, empowering individuals and groups through sustainable entrepreneurship activities. Kennedy assures readers that he is ready to collaborate. His farm has ample land to accommodate travelers and to co-create new demonstration projects and experimental structures using bricks and other materials. In the meantime, he will continue to engage with UB students through virtual projects and study abroad, and will welcome Chris and Hannah back this summer. Kennedy points out that change happens *pole pole* (slowly by slowly, in Swahili) but that we should never give up hope or get discouraged as there is so much we can do together through Experiential Learning.

11 Youth Engagement in Nakivale Refugee Settlement, Uganda

Written with Mbilizi Kalombo, Executive Director of KBTN

KBTN is a refugee-led organization in Nakivale Refugee Settlement in Isingiro District, Southwestern Uganda. KBTN stands for KYETE BIINGI TAI NYEME which translates loosely as "a shared load is always light" in one of the Congolese native languages. KBTN was founded by Baraka Benedict and Mbilizi Kalombo to empower the vulnerable communities through livelihood and environmental conservation activities. Its vision is to create resilient, sustainable, and independent vulnerable communities.

To appreciate the mission of KBTN, one needs to get close to the journeys of its founders. Benedict and his family spent over 14 years in Kyaka II Refugee Settlement in Kyegegwa District, Uganda. Like many refugee children, he was not able to attend secondary school after completing Universal Primary Education. His family simply didn't have the resources. But Benedict had a passion and a desire to change the standards of living for his family members, being the first born. He realized that only education could help him achieve this goal so he began to burn charcoal during the school holidays in order to get money for school fees. He made the charcoal from trees in Kyaka II and sold it in nearby Kampala. Charcoal is a prized cooking fuel throughout Uganda and much of Sub-Saharan Africa. Although it contributes heavily to deforestation and health problems for those producing and cooking with it, it remains a common but destructive means of revenue generation and livelihood.

After completing his Universal Secondary Education (USE)/ Uganda Certificate of Education (UCE), Benedict established a nursery school to support other refugee children in reaching their dreams. He started with a class of 15, teaching under a shade tree. Two years

DOI: 10.4324/9781003489337-14

later he raised enough money to establish two temporary structures under his registered Community Based Organization of YIDA. Shortly after this milestone, Benedict and his family were resettled to Australia by UNHCR, The UN Refugee Agency, where he would continue his education through professional courses. Benedict is now completing a Bachelor's degree from Latrobe University with plans to eventually earn a PhD since he is now capable of paying his school dues. While Benedict was determined to support his own development, he knew that other refugees were not as lucky as he was, and he was committed to continuing his humanitarian efforts with a focus on youth. In 2019 Benedict approached his cousin Mbilizi Kalombo who was living in Nakivale Refugee Settlement, some 200 km from Kyaka II, with the idea of starting a new refugee-led community-based organization. Benedict would be the CEO and Kalombo the Executive Director, and together they would form what would become KBTN.

At the time, Kalombo was in his first year of university, one of the lucky few refugees who received a scholarship to continue beyond secondary school due to his outstanding academic performance in the Uganda Advanced Certificate of Education (UACE). Kalombo had come to Nakivale Refugee Settlement when he was nine years old, seeking refuge with his family who were forced to flee the Democratic Republic of Congo. He was told that his father had inherited a hill of gold from his grandparents who were the gold miners for the rest of their lives in Kamituga, South Kivu, province that attracted politicians and others seeking to profit from its riches. His family was eventually driven from the place through violence to release their hold on the valuable property, and they made their way to Nakivale. Kalombo completed his primary and secondary schooling and then a Bachelor's degree at Victoria University in Kampala. When Benedict approached him with the idea for KBTN he was eager to help start and lead the organization.

In many ways, Nakivale was an ideal place for KBTN and its mission of supporting youth. Nakivale is the oldest refugee camp in Africa, and one of the largest in the world. Situated around 300 miles from the capital of Kampala, it is enormous in size, stretching beyond 180 kilometers, and housing over 150,000 refugees, mostly from the Democratic Republic of Congo, Burundi, Somalia, Rwanda, Ethiopia, and Eritrea. The Camp has been called the most progressive refugee settlement in the world. It is well organized and divided into 79 villages with an average of 800 to 1,000 people per village. In

Nakivale, residents live and work within three distinct zones: Base Camp, Juru, and Rubondo. As refugees settle, they are given land to farm. The idea behind the settlement is one of self-sufficiency.

Despite progressive resources and organization, however, life is not easy in the settlement, especially for youth. Some 35% of Nakivale's residents are school aged but only 10% attend secondary school and 23% of the population are between 15 and 24, almost 34,000 youth. Vocational training is weak, and youth rely on farming or other small-scale businesses to support themselves. Agriculture remains the primary economic activity but is getting harder to sustain. Allotted plots have gotten smaller, and there is increasing competition for land. And the effects of climate change are being felt throughout the settlement. The rains have been unpredictable with long periods of drought and the ground more barren. It has become more difficult to raise crops and support livestock. Lack of seeds, particularly the hardy types, and low levels of irrigation and knowledge are causing frustration and the behavior of youth is becoming more problematic. These conditions are making it difficult for young people in Nakivale and across refugee camps to get ahead. Yet the number of displaced youth across the world is increasing, because of war and conflict but also climate change. While youth should be the answer to a family's growth and prosperity, they represent a destabilizer, causing burden to the land and resources.

For Benedict and Kalombo, the vision of KBTN is personal. Benedict had a chance to leave. Although his family could not afford to support his education, he found a way forward – like many of the diaspora community, he wanted to give back. Kalombo was one of the few (only 3%) to get a scholarship to attend higher education. He was well aware of the inequalities and the need for support. On February 15, 2021 KBTN was officially registered as a refugee-led community-based organization by Isingiro District local government. It was supported through private donations that Kalombo received in Australia through fundraisers organized by Baraka Benedict with other diaspora communities. In 2023 KBTN was among the first cohort to win a refugee-led Innovation grant from the UNHCR (the UN Refugee Agency) Innovation service, a program designed to improve the lives of displaced people and communities. Funding supported their Seed Program and helped establish their work with youth within the settlement with a focus on sustainable agriculture and livelihoods. With the funding, they managed to plant 40,000 fruit trees and established three community gardens across the settlement's zones (Base Camp, Juru, and Rubondo). They supported farmers with more than 10 tons of improved seeds, agricultural tools and equipment, and farmer capacity building on different modern agriculture techniques.

When approached with the opportunity to offer a Project Challenge through the University at Buffalo ELN, Benedict and Kalombo wanted to get it right. KBTN was still a young organization and they were eager to build on the success of the Innovation grant. While there was great need across the settlement, they wanted a model that was strategic and sustainable, able to support both the youth and the fragile environment. Unsure how to best move forward, Kalombo and Benedict decided to start with a survey, interviewing the youth groups that lived and worked across Nakivale's three administrative zones. There were 150 youth groups all engaged in different agriculture and vocational projects, working to generate funds for themselves and their families. By understanding the success and needs of these efforts, KBTN could build capacity while ensuring that projects were environmentally sustainable. Since Kalombo was living and working in Nakivale, he was interested in hearing directly from the youth and introducing them to the idea of Experiential Learning. He also saw it as a perfect opportunity to collaborate and invite engagement from university students.

On Tuesday, September 12, 2023 Kalombo and his team in Nakivale began three days of data collection. They moved through the zones, completing a Google form while interviewing 150 representatives of youth groups and community leaders from throughout Nakivale. They sought to learn about vocational and agriculture activities – both current and past, and priorities for future training and resources. They knew that while Nakivale youth were industrious, not all projects were equally successful, and that critical resources were increasingly hard to come by. The survey revealed that most youth initiatives had been funded through grants or sponsorship, ending once funding ran out. The groups had difficulty maintaining the projects due to limited funds and knowledge about business and leadership. They also reported changing climate conditions that were making certain types of agriculture projects difficult to sustain. While agriculture remained a priority for youth, they wanted to diversify their businesses and develop skills that could help them build capacity beyond their time in Nakivale. They asked for training and opportunities to explore new technologies that could improve the land for farming, while also creating skills and revenue.

The survey work was pivotal for Kalombo and KBTN. In conducting the survey, they were able to connect with the various youth groups and understand their needs and priorities related to vocational training and sustainable technologies. But the survey also proved transformative for UB student Sara Bramladge. At the time, Sara was a senior Interdisciplinary Studies major interested in pursuing a job in the non-profit sector. Although Sara had already completed a Project Challenge through the ELN, she was looking to develop her assessment skills and gain experience that could lead to work opportunities following graduation. Since she had a specific interest in working with refugee populations, she was eager to work with Kalombo on compiling a report synthesizing the survey data.

Upon examining the findings of the survey, Kalombo solidified his plans to start his own version of the ELN through KBTN to be called "Nakivale Experiential Learning and Innovation Center (NELIC)." NELIC would introduce youth groups to community innovations and sustainable technologies relevant to the lives of refugees living within and outside Nakivale. He envisions hosting speakers and panelists, visiting projects and conducting interviews, and posting to a YouTube channel as a way to expose the youth groups to possible innovations. But Kalombo wants to go much further. Like ELN, he wants to support engagement and learning through PEARL digital badges, helping youth prepare through research and skill development, and connecting with mentors who can oversee projects and deliverables. Kalombo envisions digital badges that will give access to start-up funds and other critical resources while enabling the tracking of projects and project leaders, supporting dynamic research and assessment that can help strengthen the program and create a pipeline to jobs and innovation.

While Kalombo is eager to seek funding for NELIC, he is not waiting. He is putting its vision in motion through exploring sustainable technologies and engaging his own team in building capacity. Through an evolving partnership with Vetiver Without Borders-Canada, Kalombo is exploring the potential impacts of vetiver grass as a catalyst for soil revitalization and ecological restoration. As he and his team engage in initial planting, targeting focal areas that are highly degraded, they are working to build their own skills and competencies. In consultation with Dr. Mike Jabot (co-author), they are learning to test the soil using simple techniques and materials, gathering baseline data that will be mapped using available climate-science data. The team is excited to do this work. It will allow them to study impacts but also explore patterns of degradation over time to understand what has happened to the land and the soil, and to use this information to optimize their efforts. Meanwhile, Kalombo's partner, Vetiver Without Borders – Canada, is excited to develop a process for testing and geotagging that will help spread the technology to other areas that are highly degraded and in need of restoration.

As Kalombo and his team build their own capacity to engage with local partners, they will begin to develop a portfolio of Project Challenges for Nakivale youth and related resources and support. Kalombo envisions the earning of digital badges as a gateway to funding resources and mentorship while also supporting assessment. He also recognizes the opportunity to scale the model beyond Nakivale, to connect with other refugee settlements throughout Africa and beyond. While youth will surely benefit through the creation of jobs and business models, Kalombo points out that NELIC will also help transform degraded ecosystems into catalysts for regeneration and capacity building. As Kalombo looks at the future he sees nothing but opportunities. He sees Experiential Learning as a path to a much brighter tomorrow.

12 Transforming Basic Education in Enugu Nigeria

Written with Professor Ndubueze L. Mbah, Commissioner of Education Enugu Nigeria and Associate Professor Department of History, UB

Basic Education is foundational to sustainable development, but despite significant investments, it is in crisis, driving human capital underdevelopment in the Global South. Africa has an 89% learning poverty rate compared to 45% and 14% for East Asia, and Europe respectively (Bill & Melinda Gates Foundation, 2023). Nigeria, Africa's most populous country, has a 92% average deprivation rate with 48% of students dropping out before completing compulsory Basic Education (Nigerian Economic Summit Group, 2022). Without foundational skills needed for employment or job creation, youth are unable to contribute to the fragile economy. Rather than catalysts for growth and innovation, they become destabilizers, stressing vulnerable ecosystems, and perpetuating the cycles of poverty and underdevelopment.

For Nigeria and much of Africa, Basic Education represents both an urgent challenge and a powerful solution for economic and sustainable development. By improving academic outcomes and cultivating critical skills and competencies aligned with human capital needs, Basic Education can feed vocational and tertiary education systems, creating a dynamic pipeline for growth and capacity building. Since work-related skills and competencies are best cultivated through applied experiences rather than traditional curricula that prioritize rote memorization, competence-based education models, including Experiential Learning, are being embraced as promising solutions. Yet these approaches have yet to be tested within low-resource environments, or leveraged as catalysts for economic and sustainable development; that is, until now.

DOI: 10.4324/9781003489337-15

In Enugu State, in Southeastern Nigeria, a bold re-imagining of Basic Education is underway. Through the construction of 260 new Smart Green Schools, students will engage in hands-on learning connected with the SDGs and local solutions to global challenges. The reform initiative is unifying early childhood with primary, and junior secondary school levels of education within integrated campuses, creating 21st century learning environments equipped with modern Information and Communication Technology including internet connectivity, ICT and science laboratories, virtual and augmented reality, multimedia libraries, digital smart boards, and tablets for teachers and students. The Smart Schools are being constructed with low-carbon and recycled building materials and powered by clean and renewable energy. Resourced with integrated smart farms to support School Meal Plus Programs, promote sustainable agriculture, and foster circular economies of climate action, as well as child-healthcare facilities, the Smart Green Schools represent radical interventions in child welfarism and skill-oriented experiential education in Africa.

The Smart Green School Initiative includes historic investment in Basic Education and Experiential Learning, one that is directly tied to expectations for economic and workforce development. Enugu Governor, Dr. Peter Ndubuisi Mbah, views the creation of opportunities for productive employment for youth and women as the primary mechanism for lifting people out of extreme poverty and building prosperous societies. Accordingly, Enugu is embracing Experiential Learning as a vehicle for transforming youth into strategic assets for economic growth and social progress.

The Enugu Experiential Learning Curriculum is designed with this vision in mind, along with the need for a scalable model that supports teaching and learning within the Nigerian context. Because Experiential Learning is fundamentally different from traditional instructional approaches that feature passive learning, there is a need to help teachers develop their own pedagogical practice while working to introduce the model to students and key stakeholders. Accordingly, the Enugu Experiential Learning Curriculum includes frameworks for both teachers and students, CASE and PEARL respectively. These frameworks were developed in collaboration with Dr. Mara Huber (author) through translating the ELN model to meet the unique needs and opportunities associated with the Enugu Smart School initiative.

For teachers, CASE guides the development and integration of Experiential Learning across grade levels and subject areas. CASE refers to four key elements to be included in signature lessons and activities: (1) Collaboration; (2) Achievement and Adding Value; (3) Skill development; and (4) Experience and Evidence of Engagement. CASE is being used to develop the "Scheme of Work," helping pilot teachers and curriculum specialists translate the expanded curriculum (25 subject areas) into lesson plans that include demonstration activities, experiences, and project-based learning. CASE is highly flexible and is being used to generate diverse experiential activities and lessons. Teachers will have structured opportunities to share their CASE work, supporting research and assessment while building communities of practice, showcasing teaching innovation, and scaling best practices and expertise across the State.

While critical, enhanced instruction alone will not resolve Enugu's learning crisis. Enugu students must do their part to leverage instructional experiences and connect them with their own academic and professional goals. Experiential Learning calls on students to actively engage with experiences, challenging their assumptions and creating new knowledge through reflection, integration and discovery. While instructors and mentors will play a critical facilitative role, students must develop their own capacity to activate opportunities and take ownership of their learning. Accordingly, while teachers are implementing CASE lessons and activities, they will introduce the PEARL engagement framework, prompting students to Prepare, Engage and Add value, Reflect and Leverage their experiences toward broader impacts.

PEARL will be used primarily as a tool for students, allowing them to facilitate their own learning as they engage with challenges, ideas, and technologies. Initially supported through worksheets and teacher-led activities, experienced students will engage with PEARL modules to guide their own learning, taking on extended projects while working with community partners and external stakeholders. In this way, PEARL will provide a scaffolded facilitation framework while also supporting dynamic assessment and capacity building. By collecting and analyzing data on student growth and outcomes associated with the PEARL framework, the Experiential Learning Curriculum can be studied and optimized for both individual and collective impacts.

While the potential for transformative growth is compelling, especially with the commitment and scale of investments being made in Enugu, the benefits will not be easy to obtain or assess. The construction of the 260 Smart Schools and the introduction of the

Experiential Learning Curriculum represent massive disruption to traditional paradigms and structures, and the official metrics used to evaluate educational success will be insensitive to the new outcomes and expectations being cultivated. With this in mind, it will be critical to focus strategic attention on the intended outputs of the model, engaging key stakeholders to develop powerful assessment tools and dashboards that can guide iteration and optimization – supporting and ensuring success of the individual components and sustainability of the dynamic system.

To support and integrate the proposed collaboration and planning work, a Center for Experiential Learning and Innovation (CELI) is being created to bridge Basic Education and economic development priorities and investments. CELI will emerge as a dynamic mechanism for guiding, monitoring and sustaining benefits associated with a full-scale Experiential Learning Curriculum implemented across the State. In addition to leading research and assessment activities, CELI will put the Experiential Learning system in motion through the creation of SDG Innovation Challenges that will invite teachers and students to engage through mentored projects and digital badges. Innovation Challenges will be framed around Enugu's strategic priorities for growth and development which include: Agriculture, Energy and Mineral Resource Development, Commerce and Business, Urban and Rural infrastructure, Creative Industries and Tourism, and Information & Communication Technology. The Center will integrate these priorities with the Sustainable Development Goals and breakthrough technologies, inviting students to explore challenges, innovations and emerging career paths while developing related skills and competencies through digital badges and micro-credentials. In addition to Smart School students and teachers, CELI will incentivize engagement from tertiary and vocational education sectors, creating a dynamic pipeline of talent and innovation, while also connecting external stakeholders to co-create related degree programs, specializations, micro-credentials and other offerings that are mutually beneficial and tied to strategic priorities.

While exciting, the idea of Innovation Challenges is new and difficult to grasp. Pilot teachers are working on early examples that demonstrate the potential for engaging students along with key stakeholders and community organizations. Together, they have identified clean water and sanitation as a pilot Innovation Challenge. The need for accessible water is driving bold investments in Enugu to address decades-old challenges that have impacted development and undermined growth. The problems are complex and pose significant challenges to the well-being and livelihoods of residents. Factors such as the replacement of damaged asbestos pipes, the region's topography, and the presence of coal deposits have contributed to water scarcity, forcing communities to rely on inadequate water sources. Through the Water and Sanitation Innovation Challenge students are being invited to explore Enugu's needs through the lens of the Sustainable Development Goals while working toward innovation solutions.

In addition to engaging Enugu students and teachers, the Water Innovation Challenges is providing a transformative opportunity for a UB Mechanical Engineering student, Muhammed. Muhammed is from Bangladesh, a country in the Northeastern part of South Asia that experiences its own sustainability challenges, with 40% of its population lacking access to safe water. Through his own Project Challenge, Muhammed is testing a water sample recently brought to Buffalo from a river adjacent to the Enugu school. Based on his findings, he will work with Enugu pilot teachers to develop educational resources focused on water pollution, purification techniques, and creative problem-solving strategies. Through recorded lessons and Zoom sessions, Muhammed will help challenge Enugu students to develop their own filtration solutions, using local materials and working to meet formalized purification standards. He is excited to see what students come up with and is hopeful that new solutions and ideas will be generated. When asked about his ELN Project, Muhammed feels grateful for the opportunity to be part of Enugu's Experiential Learning journey and is excited to see how the students and community will benefit.

Meanwhile, Enugu teachers are busy developing additional Innovation Challenges focused on Agro-Allied Projects, Technology, Sanitation, and Renewable Energy. They have plans to plant gardens, experiment with drones, and engage the school community in producing soap and other supplies that can promote health and hygiene while generating funds to support evolving innovation challenges. As the Smart Green School model begins to scale with new teachers and students being introduced to the Experiential Learning Curriculum, Innovation Challenges will take on a life of their own, providing opportunities for new lessons, demonstration projects, and digital badges that will showcase student achievement and high-value skills and competencies.

Although still in its nascent stages, the Enugu Experiential Learning Curriculum is already garnering international attention. The same stakeholders who have been at the table to address the knowledge crisis are eager to invest in a most promising solution. And while the re-imagining of Basic Education in Enugu is just beginning, the impacts of Experiential Learning are already taking root. When asked to brainstorm slogans, the pilot teachers are quick to suggest that Enugu students are "Luminary Learners," "Curious Minds," and "Pupil Pioneers," and that Smart Schools are cultivators of "Green Dreams and Bright Futures." But perhaps most impressive are the pilot teachers themselves. While the world watches, they are boldly re-envisioning Basic Education, inspiring students to explore and engage; to stretch their understanding and contribute to sustainable solutions. Like their students, the pilot teachers are developing their own capacity through Experiential Learning. And as their practice grows and strengthens, they will be empowered to share and offer support to those who are just getting started. This is how Experiential Learning will scale. This is how transformation happens.

References

Bill & Melinda Gates Foundation (2023). 2023 goalkeepers report. www.gatesfoundation.org/goalkeepers/downloads/2023-report/2023-goalkeepers-report_en.pdf

Nigeria Economic Summit Group (2022). Nigeria Economic Summit Group session 28, (NESG 28), in Abuja.

13 Featured Partners and Their Engagement with the SDGs

Although we encourage readers to identify their own partners, building on existing relationships or cultivating new collaborations, we are particularly proud of our Global Partner Network. The following profiles have been shared by featured partners who are eager to connect with educators and students from around the world.

Featured partner profiles reference specific SDGs of focus as indicated below.

1. No Poverty;
2. Zero Hunger;
3. Good Health and Well-being;
4. Quality Education;
5. Gender Equality;
6. Clean Water and Sanitation;
7. Affordable and Clean Energy;
8. Decent Work and Economic Growth;
9. Industry, Innovation and Infrastructure;
10. Reduced Inequalities;
11. Sustainable Cities and Communities;
12. Responsible Consumption and Production;
13. Climate Action;
14. Life Below Water;
15. Life on Land;
16. Peace, Justice and Strong Institutions;
17. Partnerships for the Goals.

DOI: 10.4324/9781003489337-16

Germany

Organization: **Youth-Leader**
Location: Berlin, Germany
Vision: Omnipresent high impact youth leadership as cultural asset nurturing a collaborative, peaceful, regenerative society of able active citizens learning humanity's finest solutions and implement them in their home region.
Founder: Eric Nicolas Schneider

Youth-Leader was founded in 2005 to support the United Nations Decade of Education for Sustainable Development 2005–14. Solutions stories, hero poster sets, presentation kits and speakers got used in classrooms, at events and on expeditions on all seven continents. Today, our global community of changemakers, educators, and project partners equip youth and adults with a simple, powerful model to take simple, swift online and local actions impacting lives and land near and far, and activating an enabling environment for youth-led projects in their region. With over 400 missions and PEARL badges, YL supports Experiential Learning in classrooms and student clubs, also beyond school grounds, hours and years. By design, our model enables rich co-benefits for teachers, parents, mayors, journalists, civic groups and more.

Honduras

Organization: **Roatan Institute for Marine Sciences (RIMS)**
Location: Bay Islands, Honduras
Vision: Preserve Roatan's marine and coastal ecosystems through education and research.
SDGs: 1, 4, 6, 7, 8, 13, 14, and 15
Founder: Don Julio Galindo

The Roatan Institute for Marine Sciences (RIMS) was established in 1989 to leverage education and research to protect Roatan's natural resources while providing sustainable economic opportunities for its islanders. RIMS supports several local and international organizations in conservation efforts, including developing the largest privately-owned solar installation in the Western Caribbean, with an annual energy production of 534,000 kWh and 378 metric tons of CO_2

reduction. RIMS has led the charge to eliminate single-use plastics and leveraging biomicrobics wastewater treatment to reduce environmental pollutants that threaten over 30 miles of fringing and barrier reefs, seagrass beds, mangroves, and shorelines. RIMS works with several university groups that visit each year and conducts reef restoration along with dolphin research and training programs. The RIMS dolphin program is not just about training dolphins; it's about enabling students and researchers to collect and interpret data on behavioral signaling and social alliances within a dolphin pod.

Nigeria

Organization: **Ayomi Arts**
Location: Ibadan, Nigeria
Vision: To be the leading sustainable and innovative creative arts brand in Africa.
SDGs: 8, 10, 11, 12, 13, and 17
Founder: Ayomitomiwa Ogunsile

Ayomi Arts is a social enterprise in the creative industry incorporating art and culture with innovative strategies to repurpose waste materials while proffering solutions to the environmental impact of wastes, particularly from the creative industry. Founder Ayomitomiwa Ogunsile has over seven years of experience in the creative, environment, and real estate industry. She is a Local Pathways Fellow of the United Nations Sustainable Development Network and an Orange Knowledge Scholar. She has her first degree in Architecture and is currently studying for a Master's degree in Urban Management and Development at the Institute for Housing and Urban Development Studies, Erasmus University Rotterdam, Netherlands.

Organization: **Future Leaders Empowerment Network (FLEN)**
Location: Ibadan, Nigeria
Vision: A world where every adolescent and young adult lives an empowered, healthy, and dignified life.
SDGs: 3, 4, 10, and 17
Founder: Gideon Adeniyi

FLEN is providing a platform for every adolescent and young adult to fulfill their potential and vision through capacity building, collaboration, and social innovation. The organization has also carried out over 15 seminars, conferences, and workshops across several underserved communities in Nigeria while also distributing academic resources including books.

Gideon is a fellow of Young African Leaders Initiative Regional Leadership Center Accra Ghana and a Commonwealth Scholar with a MPH from the University of Manchester, UK. Gideon alongside Dr. Martins, PriHEMAC Executive Director were the mentors who facilitated and guided students during their ELN projects with PriHEMAC.

Organization: **Primary Health Care and Health Management Centre (PriHEMAC)**
Location: Osun State, Nigeria
Vision: Promoting the health status of vulnerable members of the community particularly the children, mothers, and elderly.
SDGs: 2, 3, 10, 11, and 17
Founder: Dr. Martins Ogundeji

PriHEMAC has implemented various projects focused on malaria, TB, sickle cell disease, maternal and child mortality among others with development partners including Physicians for Social Responsibility, United Nations Children Fund and Global Fund. It is, however, the negligence by government and development partners and lack of programs for older persons that led PriHEMAC to start the program on promoting elderly friendliness through empowered stakeholders. We have collaborated with the ELN on this program since the COVID-19 pandemic. The founder is a public health consultant who retired as the Deputy Director at National Primary Health Care Development Agency.

Rwanda

Organization: **Kigali Genocide Memorial**
Location: Kigali, Rwanda

The Kigali Genocide Memorial is the result of a collaboration between Rwandan authorities and the Aegis Trust for Genocide Prevention, born from the National Holocaust Centre in the UK. Opened in 2004, the Memorial became the starting point for peace and values education now built into Rwanda's national school's curriculum. Strengthening community resilience against division, it is ripe for adaptation and use internationally. If peace can be built after the Genocide against the Tutsi, it can be built anywhere.

Organization: **Urukundo Learning Center and School**
Location: Muhanga, Rwanda

Hundreds of thousands of children who lived through the 1994 genocide and its aftermath in Rwanda struggled for survival in desperately impoverished situations. From this devastating time of death and destruction Urukundo was created to reach out to these vulnerable children. What began as a home for girls, quickly grew to include boys and infants. By the tenth anniversary, Urukundo purchased land and built two homes, one for the boys and one for the girls. Out of necessity, they started a farm to supply vegetables, milk, eggs, and meat. Recognizing the need for quality education Urukundo opened a school (preschool-Primary 6) which now serves over 1,200 students and has received National School of Excellence designation by the Rwanda Ministry of Education. All Urukundo Primary (elementary) school graduates go on to attend schools of higher education. Professions of Urukundo graduates include lawyers, nurses, IT specialists, librarians, teachers, chefs, and business administrators. Urukundo Learning Center is located in the district of Muhanga which is an official International Sister City of Buffalo, NY. The Kinyarwanda word "Urukundo" means "Love."

Tanzania

Organization: **Community Life Amelioration Organization (CLAO)**
Location: Mwanza Tanzania
Vision: To see a sustainable community where everyone has a decent and contented life.
SDGs: 1, 2, 3, 4, and 5
Founder: Kennedy Mahili

Community Life Amelioration is a registered nonprofit organization that fosters utilization of skill and manpower for sustainable development in both social and economic sectors and supports vulnerable communities in developing skills toward revenue generation and empowerment. CLAO provides training and support for community technologies including water filters, WASH education, soap making, bee keeping, tailoring, brick making, fero cement water catchment tanks, perma garden farming, and other projects that focus on sustainability and empowerment. CLAO prioritizes out of school youth, people with albinism, widows and other marginalized communities without access formal education. CLAO has engaged extensively with the ELN and UB students and serves as a host for study abroad courses, partnering with students and faculty on collaborative construction projects using soil-stabilized bricks.

Organization: **Community Volunteer Services Tanzania (CVS-Tanzania)**
Location: Ibadan, Tanga, Tanzania
Vision: Having a youth society who are able to speak out, defend and champion for their sexual and reproductive health and rights in Tanzania.
SDGs: 3, 5, and 6
Founder: Simon A. Mashauri

Community Volunteer Services Tanzania (CVS-Tanzania) is an organization located in the Tanga region of Tanzania with the mission of enhancing the sexual and reproductive health of adolescent girls and young women by establishing, strengthening, and engaging youth clubs both within and outside of schools. Our strategy focuses on transforming the attitudes and behaviors of young individuals towards a healthier sexual lifestyle, aiming to prevent unintended pregnancies, abortions, sexually transmitted infections, including HIV/AIDS, as well as addressing family planning and other sexual and reproductive health challenges. We welcome global interns and volunteers and provide students with invaluable experience to advance their careers while contributing to our mission.

South Sudan

Organization: **Abukloi Foundation**
Location: South Sudan
Vision: A peaceful, healthy, educated, and economically self-sustaining South Sudan.

Mission: Promotion of sustainable development of local communities through capacity building and participatory community-driven initiatives.
SDGs: 2, 3, 4 and 13
Founder: Angelo Maker, a lost boy of Sudan.

"Abukloi" is a Dinka word meaning "We Can Do It." The name reflects the vision, commitment, and determination to work together with the people of this region to bring about meaningful progress through the education of future leaders and the creation of a variety of self-sustaining projects. The country of South Sudan was created in 2011 and was born out of the resolution of decades of war in Sudan. The staggering conflicts in the region brought Sudan and its people to the attention of the world and the stories of thousands of children who fled Sudan by traveling on foot over hundreds of miles to reach safety in nearby African countries. One of these many Lost Boys is Angelo Mangar Maker, who provided the vision for Abukloi Foundation. Abukloi Foundation is currently, operating in the following areas: Education; Agriculture; Health; Climate resilience; and Economic and Community Development.

United States

Organization: **The Collaborative Center for Social Innovation**
Location: Buffalo, NY
Mission: Empower youth to better our community by fostering empathy and instilling a sense of purpose and belonging.
SDGs: 3, 4, 11, 13, 14, 15 and 16
Founder: Apryle Schneeberger

The Collaborative Center is a 501(c)(3) nonprofit dedicated to providing teens with innovative experiential learning opportunities to build transdisciplinary academic knowledge, social emotional skills, and professional networks as they develop, share, implement, and iterate community-driven and data-informed solutions to real time sustainability challenges. Emphasis is placed on sustainability challenges centered on the social determinants of health, health equity, and climate resilience. As part of this work, The Collaborative Center engages high school students in challenges in the ELN Project Portal.

Organization: **SUNY COIL Center**
Location: New York, USA
Mission: To cultivate an engaged educational environment by providing the resources, support, and professional network to enable educators and institutions to incorporate virtual engagement opportunities for all students and faculty.
Founder: Jon Rubin

Collaborative Online International Learning (COIL) connects students and professors in different countries for collaborative projects and discussions as part of their coursework. COIL Collaborations between students and professors provide meaningful, significant opportunities for global experiences built into programs of study. COIL enhances intercultural student team focused interaction through proven approaches to meaningful online and virtual engagement, while providing universities with a cost-effective way to ensure that their students are globally engaged. The SUNY COIL Center pioneered the COIL model in the early 2000s, and has been at the forefront of empowering professors, students, programs and institutions to embrace diversity through inclusive teaching and learning focused on equity while connecting through difference.

Uganda

Organization name: **Africa Diabetes Alliance**
Location: Kampala, Uganda
Mission: The mission of A*f*DA is to effectively and sustainably empower people, organizations, and systems to disarm, and ultimately end the diabetes epidemic in Africa.
SDGs: 1, 2, 3, 4, 5, 8 and 17
Co-founder: Edith Mukantwari (BFST, MPH)

From its inception in 2018, Africa Diabetes Alliance (A*f*DA) has focused on leveraging research and training opportunities in order to drive the patients' and caregivers' understanding of diabetes to improve self-management for improved health outcomes and quality of life with great success. This involves the organizing of annual diabetes education events, sharing digital resources on a WhatsApp support group where questions are answered and discussions are had as well as taking on global, regional and local research and advocacy events. In 2023, A*f*DA expanded focus areas to include health systems strengthening support as well as a social enterprise that supports financial growth and sustainability.

Organization: **Batwa Pygmies Network**
Location: Kanungu Uganda
Mission: The Batwa Pygmies Network is an indigenous-led international organization dedicated to fostering a brighter future for the Batwa Pygmies, who are among Africa's original indigenous people of the forest.
SDGs: 1, 10, and 16
Founder: Isaiah Wycliff

The global population of Batwa people includes 70,000 to 87,000, spanning Uganda, Rwanda, Burundi, and the Democratic Republic of the Congo (DRC). Our mission is to serve as a voice for our communities worldwide, advocating for their rights, well-being, and cultural preservation. Key initiatives include: Sharing biodiversity knowledge, highlighting the challenges we face in the 21st century including questions about our survival as a people, and learning from other indigenous communities. The paradox of being displaced from our ancestral lands to save gorillas is a pivotal discussion we aim to engage in. Together with gorillas and the global ecosystem, we can create a sustainable community. Join us in our journey to protect our heritage, advocate for our rights, and build bridges across cultures. Together, we can create positive change for the Batwa Pygmies and the world.

Organization: **Biringo Women's Development Association (BIWODA)**
Location: Kabale District, Uganda
Mission: BIWODA was formed as a collective of women farmers from Biringo Uganda with the goal of working together to address community development and achieve self-sufficiency.
SDGs: 4, 5 and 7
Founder: Deo Mbabazi

BIWODA is a rural women's group established in 1988 but formalized in 2014. With a current membership of approximately 180 rural women, BIWODA aims to empower its members by improving living standards through sustainable organic farming, skills development, education, and clean energy innovations. Located in Buramba parish, Kahungye- Rubaya Sub County, Kabale District, Uganda, the association operates in an area marked by limited social services and development opportunities. Recognizing the role of women as primary breadwinners in their families, BIWODA seeks to address the income disparity by engaging in various income-generating activities including goat and poultry rearing and organic farming. Since its legal formation, BIWODA has forged partnerships with government programs, local community-based organizations (CBOs), NGOs, and international entities like University at Buffalo and Bees Abroad.

Organization: **Gender Tech Initiative-Uganda**
Location: Gulu City, Uganda
Vision: To create a community where women and girls are fully empowered to contribute to and lead in the innovation and progress of science and technology, and where their voices and perspectives are valued and heard.
SDGs: 4, 5, and 9
Founder: Ruth Atim

The Gender Tech Initiative-Uganda was founded in 2017. Our mission is to empower women and young girls to embrace science and technology. Readers are invited to check out the GenderTech podcast on YouTube which is aimed at building digital skills and providing support networks for women and girls navigating digital spaces. We also have in-person events where we actively engage with young children and youth through interactive activities such as coding camps and STEM-focused competitions, aimed at igniting their interest in technology and fostering a passion for innovation. These learning opportunities equip the young with practical skills and also empower them to become leaders and change-makers in their communities who drive progress towards gender equality in STEM fields and beyond.

Organization: **Kyete Biingi Tai Nyeme-KBTN**
Location: Nakivale Refugee Settlement, Uganda
Vision: To create resilient, self-sustainable and independent vulnerable communities.
SDGs: 2, 4, 13, and 15
Founders: Mbilizi Kalombo and Barak Benedict

Nakivale Refugee Settlement faces critical challenges of land degradation, diminished agricultural yields, and environmental vulnerabilities. Traditional farming practices and climate change are exacerbating these issues, threatening the community's livelihood and food security. KBTN, with the support from UNHCR, has so far distributed quality seeds and provided training in agronomic practices to 555 households and supported 5 VSLAs (Village Savings and Loans Associations) with agricultural tools and equipment and revolving funds. Also, it has established community gardens and planted up to 40,000 fruit trees. Nakivale Experiential Learning and Innovation Center (NELIC) will promote and support engagement around Climate Action with a focus on Environmental Conservation, Agroforestry and Community Technology. NELIC will support youth groups while serving as a model for other rural and low-resource environments where youth are eager to make change while developing their own skills and competencies and earning revenue to support themselves and families.

Organization: **Mic Gum**
Location: Gulu, Uganda
Vision: An empowered and independent arts community.
SDG: 8, 10, and11
Founder: Ayenyo Joanita

Mic Gum (an Acoli phrase which loosely translates as Talent is Gift) is a talent agency that scouts and nurtures talent from within the community, and showcases it to the world with the hope that the artists can leverage talent in future. Currently, Mic Gum focuses on writing and singing, and some of their work with singers can be found on YouTube under Mic Gum.

The arts have always been undermined in our community and often associated with failure or those who have failed in life, a narrative Mic Gum is working hard to change. Mic Gum is open for global collaboration with likeminded individuals or companies who share the same vision or wish to support the arts.

Organization: **Solar Health Uganda (SHU)**
Location: Uganda
Vision: Safe energy for all, with a priority on vulnerable women and children.

SDGs: 1, 3, 4, 7, 13
Founding Consultants: George Mike Luberenga, Caroline Mwebaza, and Siriman Kiryowa.

Solar Health Uganda (SHU), established in 2017, addresses energy poverty and climate change through collaborations with local and international partners. Notable achievements include distributing over 30,000 Pico solar lights and electrifying 86 clinics in Uganda's prioritized districts, sponsored by Let There Be Light International. Collaborating with Let There Be Light International, SHU has co-designed innovative clean energy programs like Safe Births + Healthy Homes, Lights 4 Learning, and the Shine On Network. Through surveys and outreach, SHU identifies and supports communities affected by energy poverty, electrifying rural health clinics, and distributing solar lights to vulnerable off-grid rural homes. SHU raises awareness and partners with organizations to enhance health, education, and environmental outcomes. In May 2023, SHU consultants launched the Small Village Clean Gardening Initiative Uganda, promoting food security through solar-powered irrigation and plastic waste management, prioritizing drought-affected villages, and supporting smallholder farmers.

Organization: **Support the Diabetics Organization (SUTEDO), Uganda**
Location: Gulu, Uganda
Vision: Arming all Ugandans young and old, rich and poor in the battle against diabetes as we envision a diabetic managed community with reduced prevalence.
SDGs: 3, 4, 8, and 10
Founder: Joanita Ayenyo

SUTEDO has been grounded in diabetes care since 2015. Since its inception thousands have benefited from free services including diabetes screening and education for people living with diabetes and their caregivers to manage the condition. SUTEDO runs diabetes awareness campaigns with a flagship event to highlight Diabetes type 1 which affects children and young people and often goes undetected. SUTEDO also provides direct support to healthcare facilities to improve treatment and prevention. SUTEDO has been able to succeed because of the power of collaboration and is seeking new collaborations and partnerships.

Organization: **Umbrella for Journalists in Kasese (UJK)**
Mission: UJK empowers and advocates for rural journalists in Uganda through training, legal defense and legislative lobbying to ensure independent and trustful journalism.
Founder: Bikeke Saimon
Location: Kasese District, Uganda

UJK an indigenous organization that is working towards enhancing freedom of expression, safety of journalists and amplifying community voices in the Rwenzori subregion.

Part IV

Synergies and Broader Impacts

14 STEM Education and Global Climate Change

One of the most exciting features of the Experiential Learning Network (ELN) model is that it can transcend disciplinary boundaries, leveraging Project Challenges and PEARL to connect and integrate approaches. This potential aligns with the growing appreciation for the need of science to extend beyond its typical content silos, taking advantage of the insights that other science disciplines and fields of mathematics and engineering offer. In this way, STEM education and Experiential Learning represent important areas for synergy and integration.

STEM Education

As a broad overview, STEM education refers to an interdisciplinary approach to learning that integrates Science, Technology, Engineering, and Mathematics. It emphasizes hands-on learning experiences, problem-solving, critical thinking, and collaboration. As a link to the work that we have described thus far, there are several key aspects of an integrated STEM approach that warrant consideration. STEM education breaks down traditional subject boundaries and encourages students to explore connections between different disciplines. For example, to truly understand how weather systems work, there needs to be an understanding of the physics that drives energy transfer in the atmosphere as well as the mathematical models associated with the phenomena. But a deep and robust understanding of weather also must include the development of understandings that cross into the fields of biology where atmospheric conditions serve to shape the biomes that support living things including humans.

DOI: 10.4324/9781003489337-18

In this way, STEM education emphasizes the interconnectedness of domains and systems and the importance of helping students leverage complexity to achieve new insights and understanding. While STEM education offers great promise related to innovation and problem solving, it necessitates an emphasis on practical, hands-on project-based approaches and real-world problem solving activities. Similar to the Project Challenges detailed in this book, these experiences help students apply their theoretical knowledge within practical contexts and develop important skills such as creativity, innovation, and perseverance, which are included among the 21st century skills and competencies.

Prior to the growing focus on bringing science disciplines into partnership with mathematical and engineering design, educators approached learning in what can best be described as a hypothetical where real-world constraints were not considered. In a STEM education model, this type of learning is replaced with a greater focus on asking students to identify problems, formulate hypotheses, design experiments or solutions, analyze data, and draw conclusions. Similar to the ELN approach to Project Challenges, STEM education fosters critical thinking and logical reasoning, abilities that will become increasingly important for success in the future. By connecting learning to real-world applications and challenges, students can understand the practical implications of what they are learning and prepare for future careers and opportunities.

Technology represents another important point of synergy between these two approaches. In STEM education, technology is a tool for learning and also a subject for study. In the work that we have shared using Project Challenges and PEARL, technology plays a central role as well. The use of a variety of technological tools and resources to collect and analyze data, model phenomena, simulate experiments, and create prototypes, as well as to expand collaboration are important to activating the potential of Experiential Learning. In today's rapidly evolving technological landscape, STEM skills are in high demand. From artificial intelligence to biotechnology, advancements in Science, Technology, Engineering, and Mathematics are driving innovation across industries. STEM education equips individuals with the knowledge and skills needed to participate in and contribute to these advancements.

Our introductory description of what STEM education is has focused on the pedagogical practices and shifts that are beginning to transform classrooms. But STEM education also offers important overarching impacts that reach past science to include mathematics and engineering disciplines. Like Project Challenges that are connected to the Sustainable Development Goals, STEM approaches also have societal impacts. It is a sad commentary that the STEM disciplines have had a longstanding challenge around equity and inclusion of all. However, there have been great improvements in ensuring that all students, regardless of gender, race, socioeconomic background, or ability, have access to the high-quality STEM learning experiences that will help to shape their lives. This includes providing support and opportunities for underrepresented groups in STEM. We will expand on this in a later chapter where the construct of open science is discussed.

Overall, STEM education is designed to cultivate a new generation of thinkers and innovators who are equipped to tackle the complex challenges of the 21st century and contribute to scientific advancement, technological innovation, and economic prosperity. There is no area in greater need of STEM innovation than addressing global climate change.

Global Climate Change

Climate change is one of the most pressing issues of our time, with far-reaching implications for the planet and its inhabitants. Addressing this challenge requires a well-informed populace capable of understanding the complexities of climate science, its impacts, and potential solutions. Therefore, an effective climate education framework is essential for empowering individuals to become agents of change through climate action.

Climate change is a complex phenomenon driven by a variety of factors, both natural and human-induced. A primary driver is the release of greenhouse gases (GHGs) into the atmosphere. The most significant of these gasses are carbon dioxide (CO_2), methane (CH_4), and nitrous oxide (N_2O). Human activities such as burning fossil fuels (coal, oil, and natural gas), are major contributors as well as the deforestation of landscapes and agriculture practices. In the case of deforested landscapes, the trees and vegetation act as carbon sinks, absorbing CO_2 from the atmosphere. Deforestation, particularly in tropical regions, reduces the Earth's capacity to absorb CO_2, leading to higher concentrations of greenhouse gasses in the atmosphere. Agricultural practices often produce methane emissions as well as contributing to changes in land use, such as converting forests into agricultural land further altering local climates.

While human activities are the primary drivers of current climate change, it would be unfair to not recognize that there are also natural factors that contribute. Natural factors such as volcanic eruptions that increase the level of atmospheric aerosols, and natural cycles like El Niño and La Niña can also influence climate patterns. However, there is widespread consensus by scientists that these natural factors alone cannot account for the observed changes in climate over the past century.

One of the greatest challenges that we face in trying to address global climate change is that the climate system is driven by feedback loops. Often these feedback loops exacerbate the problem. For example, as temperatures rise, permafrost thaws, releasing methane – a potent greenhouse gas – into the atmosphere, further contributing to global warming.

The impacts of climate change can be seen in many of the case studies that we have shared. And as our partners have articulated, it is often the most vulnerable communities and regions that are impacted the most severely by compounding challenges of food security, water scarcity and public health. Many of our community partners detailed in Part III are focusing on adaptation strategies to help foster resilience to the climate impacts that are affecting their communities. The use of Project Challenges and PEARL have shown great promise in helping to shape and support this work.

At the risk of admitting defeat, we feel that we have crossed over the tipping point where our ability to end climate change has passed. We are now firmly in an age where climate mitigation and adaptation are needed. A number of positive examples have been shared where the ELN model is being used to support strategies around the transformation of communities in light of climatic impacts. What is especially impressive with these approaches is that communities are empowered to bring the local knowledge and context that they uniquely possess in addressing the challenges they are facing. The iterative nature of the ELN model allows those involved in action to be informed by continually refining their strategies based on the inputs that arise from their work as well as refocusing efforts based on results. Additionally, the work that has been done has been further informed through the use of technology and digital platforms to expand access to information and to further foster collaboration. A discussion of these technologies follows in the next chapter. The ELN model has played a crucial role in helping to equip community partners with the knowledge, skills, and motivation to help to begin to address the challenges of climate change in their communities by helping to empower community members from all walks of life to become proactive stewards of the environment and advocates for a more resilient and equitable future.

15 Geospatial Technologies and Open Science

E.O. Wilson is quoted as saying that "we are drowning in data but starving for wisdom." We live in a world where we have never had greater access to data and the technologies to help us understand the world around us and how we can address the challenges that society faces. The paradigm shifts that are occurring as we begin to take advantage of the marriage of technologies and data are striking. Amongst the most important and influential of these paradigms are those of geospatial technologies and open science/data. We will begin our investigation with an overview of geospatial technologies before circling back to look at the empowerment of people through open science/data.

Geospatial Technologies

Geospatial technologies refer to a broad range of tools and techniques used to acquire, analyze, manage, and visualize geographic and spatial data. These technologies are integral to understanding, interpreting, and managing various aspects of the Earth's surface and its features as well as investigating the changing nature of both earth systems and human interactions. In the work described in this book, we have taken advantage of geospatial technologies to empower partners in their communities to initiate change. This has been in the form of humanitarian mapping efforts, which we will describe as we move forward. Before doing so, however, it is important to broadly outline the basic components of geospatial technologies that can then be highlighted in relation to our work.

DOI: 10.4324/9781003489337-19

At the heart of geospatial science is a Geographic Information System (GIS). GIS is a powerful tool for capturing, storing, analyzing, and managing spatial and geographical data. It allows users to create maps, perform spatial analysis, and visualize data in meaningful ways. Because of the wealth of data that is available, the reach of GIS has been extended into areas such as urban planning, environmental resources, agriculture, and disaster management. In order to provide data that is usable within the GIS, global positioning systems (GPS) are deployed. GPS is the satellite-based navigation system that provides accurate positioning and timing information anywhere on Earth. GPS receivers use signals from multiple satellites to determine their precise location, enabling for the level of precision needed to make data brought into the GIS meaningful and valid.

Because of the scale of many of the systems that we seek to understand, it would not be feasible to rely solely on the collection of in-situ data through GPS receivers on the ground. Instead, we are able to take advantage of a wealth of remotely-sensed data that is available. Remote sensing involves the collection of information about the Earth's surface from a distance, typically using sensors mounted on satellites, aircraft, or drones. Remote sensing data includes imagery, elevation models, and other types of geospatial data. It can be used for monitoring land cover changes, environmental assessments, disaster response, and natural resource management.

Geospatial data is special in that it allows users to both ask and answer questions they seek to address. The use of geospatial analytics uses the unique features that spatially informed data have to offer. Geospatial analysis involves the application of statistical and computational techniques for extracting meaningful patterns, relationships, and insights from spatial data. Through spatial statistics and modeling, network analysis, and other methods for analyzing spatial phenomena, we are empowered to make informed decisions.

One of the challenges that we face when attempting to bring the power of geospatial science into play, is the relatively high level of technological expertise that is often needed. A very useful and efficient way to overcome this challenge is by using web mapping and easily accessible platforms through which the products of geospatial analyses can be shared with users. These tools can empower users to access, share, and interact with geographic information in a format that is much more accessible while also offering community members the opportunity to inform those supporting the geospatial work. There are a number of ways that these platforms provide tools for advancing these goals. These include interactive maps, the geocoding of addresses, routing, and spatial data visualization, all allowing the geospatial information to be more accessible to a wide range of users.

Geospatial technologies play a critical role in many of the examples in which Project Challenges and PEARL have been applied. We have been very fortunate for the support offered from partners as we have sought to move solutions forward. Geospatial technologies continue to evolve rapidly with advancements in satellite technology, sensor technology, data analytics, and cloud computing, offering increasingly powerful capabilities for understanding and managing the world around us and will extend the influence that the PEARL model will have as new partners and projects are brought into play.

An important consideration around geospatial science and technologies, and their role in the work described in the implementation of PEARL, is the realization that much of the world is not mapped, or at best has maps that do not accurately reflect what is occurring on the ground. This makes it difficult and sometimes impossible to assist local communities in helping to address the major challenges they are facing. We have adopted humanitarian mapping as our focal construct to bring the power of geospatial technologies to the communities with which we have partnered. The adoption of a humanitarian mapping lens allows the use of GIS to create, validate, and share maps with the local communities and partners.

In our work, mappers rely on real-time data captured by satellites and supplement this information with on the ground truthed data they collect. These data include images, videos, and qualitative reports. Through a combination of GIS and in-situ data, mappers explore important patterns such as the impact of climate change, environmental degradation as well as patterns of regeneration and restoration that occur through their efforts to improve their community. By studying these patterns with partners living and working within the community, the ELN model can help develop more effective tools and practices, customizing approaches for specific communities and contexts.

We have shared the obstacles that a highly technical approach like geospatial technologies can address. But what we find to be the most exciting about the adoption of a humanitarian mapping lens is that we have seen that almost anyone can learn and contribute to this work. The basic skills are easy to master. And our global partners are providing opportunities to map their respective regions, sharing their own pictures, videos, and reports of climate conditions impacting their communities. Our partners are heavily invested in this work and eager to benefit from the ultimate goal which is to inform ecorestoration and regenerative practices. They are both our partners for assessment and eventual implementation.

This humanitarian lens encourages collaboration with local communities to gather local knowledge, validate map data, and ensure that mapping efforts are culturally sensitive and contextually appropriate. Engaging with communities empowers them to participate in the response process, strengthens local capacities, and promotes ownership of the mapping data and tools. The crowdsourcing of this data includes tracing satellite imagery, digitizing paper maps, and adding details such as roads, buildings, and water sources to create comprehensive maps of areas which helps fill information gaps and provides valuable insights..

Open Science

We have suggested that geospatial technologies are a powerful tool and that through their utilization via Experiential Learning and PEARL, both students and community partners can benefit in transformative ways. While exciting, this assertion begs the critical question, do our partners have access, access to the tools they need, to the resources that help build their background knowledge in support of their lived experience, and to rapidly changing information that would help inform their decision making. Sadly, these questions exist well beyond the scope of the work that we are sharing. As we write, communities around the world are compromised by factors such as infrastructure, education levels, socioeconomic conditions, and government policies. The most shocking realization that the science community has had is that these constraints are institutional and designed and supported in ways that make it nearly impossible for impacted communities to overcome.

There has been an incredibly important development in recent years towards overcoming the obstacles that the current systems place on communities. The movement towards open science/data has the potential of allowing all citizens to gain access and to do so in a way that levels the playing field for all communities worldwide. The practice of open science and open data is engaged in the identification and development of science that is in service to community-identified needs and allows engagement to be accessible to everyone. These represent paradigm shifts away from the typical structure of science. Open science is of critical importance and brings numerous benefits to the scientific community, society, and the progress of knowledge. Following are several reasons why, as a field, we consider open science to be a game changer.

Open science by its nature promotes both transparency in the research process as well as its methodologies. By allowing research data, methods, and findings to be openly available to all involved, it allows for scrutiny by researchers and community partners alike to help validate these data scientifically while also informed by in-situ observations, which are in a sense the lived experience of these data. This process is critical to fostering trust and credibility between researchers and community partners around these endeavors.

Because of the adoption of this "shared scientific discovery", Open science also serves to facilitate the rapid dissemination of knowledge, which in turn serves to accelerate the pace of further discovery and innovation. Through the process of openly sharing all research outputs, those involved in the process, including scientists and community partners alike, can build on each other's work more efficiently, leading to faster progress in various fields.

The work and methodologies that we have described so far have emphasized the incredible power of locally-focused interventions. Open science has similar impacts, but the paradigm expands opportunities at a much larger scale. By its nature, open science encourages global collaboration. Because research processes and methodologies are shared openly, researchers from different parts of the world are able to access and contribute to the generation of a shared dataset and expand the analysis, insights and findings around the work being completed. The expansion of the "reach" that open science offers serves to foster a collaborative and inclusive approach to scientific inquiry that may very well be exponential in its impact.

Under current practices, lack of transparency and collaboration often lead to an overall scientific process that is inefficient at best and often exclusionary in its most extreme. In particular, as we seek to address the issues that are brought forward by the SDGs, the constraints on resource allocation become acute. The sharing of research data and resources reduces duplication of efforts. Researchers and community partners can leverage collected data and shared methodologies to innovate and deepen understanding as opposed to replicating work which often occurs when data and findings are not shared broadly or in a timely manner because of institutional constraints such as paywalls and time delays due to extensive review processes. This shift allows for both the optimization of resources as well as contributing to the development of a sustainable research ecosystem.

Open science enhances public engagement by making research accessible to a broader audience. When the public can access and understand scientific findings, it fosters trust in the scientific process and allows citizens to make informed decisions based on evidence. Open science also provides valuable educational resources. Openly accessible research materials, datasets, and publications can be used for educational purposes, benefiting all community partners. Open science practices, such as sharing research protocols, data, and code, contribute to addressing the reproducibility crisis in science. Others can verify and reproduce research findings, strengthening the robustness of scientific knowledge.

Open science aligns with ethical considerations in research. Sharing research outputs ensures that the benefits of scientific inquiry are distributed more equitably, and it promotes responsible and ethical conduct within the scientific community. Many funding agencies and institutions now emphasize open science as a requirement for research projects. Adhering to open science practices can be essential for obtaining funding and complying with institutional policies.

In summary, open science is a transformative approach that promotes collaboration, transparency, and accessibility in scientific research. It contributes to the advancement of knowledge, encourages global cooperation, and enhances the societal impact of scientific endeavors. Once we agree to the adoption of open science as our paradigm, we must ensure that open science can come into play to support those it is intended to address. In the context of open science, we as a community have adopted FAIR, an acronym that stands for the principles of Findability, Accessibility, Interoperability, and Reusability. These principles are adopted to enhance the openness, accessibility, and usability of research data and other digital scholarly objects. Components of the FAIR principles are as follows:

Findability

It is crucial that data and digital objects can be easily located by both humans and computers, that these objects also have metadata and other essential information deeply described and easily searchable, and that persistent and unique identifiers be used to ensure the findability of the data.

Accessibility

Once the data or digital object is found, it should be easily accessed. Accessibility includes providing open access to the data or, if restrictions apply, providing clear information on how to gain access. Accessibility also involves considering ethical and legal considerations regarding data sharing.

Interoperability

The data and digital objects we create should be formatted and structured in a way that allows for seamless integration with other datasets or tools. We need to use common standards and protocols which will serve to enhance interoperability and will facilitate integration into larger research ecosystems.

Reusability

As we design and document data and digital objects we need to do so in a way that makes them understandable and usable by others. We should adopt open licensing of these products and clearly establish the provenance of these products. We also need to offer the supporting documentation, including methodologies and contextual information, to aid in the interpretation and reuse of the data. The FAIR principles were developed to address the challenges associated with the growing volume of digital data and the need for increased collaboration and openness in scientific research. Implementing FAIR principles contributes to the creation of a more transparent, collaborative, and efficient research environment, allowing for the effective sharing and reuse of research data across disciplines and institutions.

Bibliography

FAIR Principles. (n.d.). GO FAIR. Retrieved April 21, 2024, from www.go-fair.org/fair-principles/

UNESCO Recommendation on Open Science – UNESCO Digital Library. (n.d.). Retrieved April 21, 2024, from https://unesdoc.unesco.org/ark:/48223/pf0000379949

16 NASA GLOBE and the Sustainable Development Goals

The NASA GLOBE (Global Learning and Observations to Benefit the Environment) Program (www.globe.gov) stands as a beacon of collaborative scientific inquiry and environmental education. Over the past 29 years since its founding in 1995, the GLOBE program has been a leader in environmental science diplomacy and has been a key contributor to environmental literacy, scientific research, and global cooperation.

The NASA GLOBE Program represents a pioneering initiative that integrates scientific research, education, and international collaboration to enhance understanding of Earth's environment. In many of the same ways the tenets of PEARL have shaped the work described in the case studies shared in Part 3, GLOBE has served to link community partners from around the world to address some of the most challenging issues facing society and the environment.

The GLOBE Program was established on Earth Day in 1995 as a joint endeavor between NASA, NOAA, and NSF and has grown into a global network of 127 countries (as of 2024) and partnerships that bring hands-on, in-situ learning to play in K-12 classrooms and within communities as an exceptional example of citizen science. It is important to define citizen science in the context of this work and the expanding role that NASA GLOBE is beginning to play across communities globally.

Citizen science plays a pivotal role in advancing scientific research by harnessing the collective power of community members and professionals alike around the process of data collection, increased spatial coverage, temporal monitoring, public engagement, and education, and the development of innovative solutions that are an

DOI: 10.4324/9781003489337-20

outgrowth of the shared work. Citizen scientists contribute to vast growing data that is available through the NASA GLOBE database. Currently, there are over 250 million data points in the NASA GLOBE database contributed from over 41,000 schools and community organizations and by over 250, 000 citizen scientists working within their communities. These data are collected across the diverse communities at a scale that would not otherwise be possible because of the normal constraints of the scientific enterprise and are serving as key data sources for monitoring biodiversity and the impacts of climate change among other community-driven initiatives.

Through the 127 countries that constitute the NASA GLOBE community this citizen science program and its projects are able to extend their reach to remote or inaccessible regions through the collection of in-situ data. The expanded coverage that the NASA GLOBE program provides fosters valuable insights into areas that might otherwise be overlooked, enhancing our understanding of the issues that we need to investigate and address.

Because the NASA GLOBE Program has been in existence for nearly three decades it allows for long-term studies and the consistent and sustained monitoring of study sites. This type of work would be nearly impossible for research teams to sustain were it not for the work of the GLOBE community and its highly engaged members. This work helps to ensure continuity and enables scientists to analyze trends and patterns over time, such as seasonal shifts, species migrations, and habitat changes.

When established, the NASA GLOBE Program focused on implementation of inquiry-based science in K-12 settings. This places NASA GLOBE at the forefront of engaged STEM education using citizen science as its driver. This focus on K-12 STEM education continues today, but a very important evolution for NASA GLOBE has risen to prominence. The introduction of the NASA GLOBE Observer platform has served to welcome community members from outside formal education settings and has done an incredible job of serving to foster public engagement with science by involving these community members in hands-on research activities. Through participation, volunteers gain a deeper appreciation for scientific inquiry and develop a greater understanding of complex issues. This engagement also promotes scientific literacy and encourages informed decision-making within communities.

As we have shared through the previous chapters in this text, by welcoming the diverse backgrounds and perspectives of participants, innovative and impactful solutions to the challenges that communities face can often arise. This is also true with the work that the NASA GLOBE Program is bringing to play in the communities involved. The crowdsourcing of ideas from a broad range of participants sparks creativity and facilitates interdisciplinary collaboration, leading to novel approaches for problem-solving and discovery. An example that exemplifies this is the GLOBE International Virtual Science Symposium (IVSS) where participants share their research and solutions to a global community. In spring 2024, 285 reports from 30 counties were shared with the GLOBE community around the focused investigation of Climate and Resilience, This is an incredible example of how a citizen science focus can serve as a powerful tool for amplifying the efforts of scientists expanding their reach, and enriching research endeavors through collaboration with partners worldwide.

NASA GLOBE and the United Nations Sustainable Development Goals

Throughout this book we have focused on how Project Challenges engaging community partners can help to address and contribute to the challenges that are identified by the United Nations Sustainable Development Goals. Because the NASA GLOBE Program is built on a citizen science framework informed by local conditions it serves well as a mechanism for communities to begin their focus on identifying how the UN Sustainable Development Goals (SDGs) come to play in their locale.

The NASA GLOBE Program uses an Earth System Science Framework to show how each of the individual earth spheres are connected and interact. In much the same way, the United Nations delineated the key challenges that we as a society face into the 17 SDGs. But in reality, these individual SDGs are interdependent on each other in much the same way as the earth spheres, where changes in one area often influence other areas. In the examples that follow, we have delineated the individual SDGs simply to focus the discussion. There is no doubt that the work described serves to inform other areas as well.

Overall, citizen science can contribute in very significant ways to advance the UN SDGs by empowering individuals and communities to participate in scientific research and contribute to addressing global challenges. Some brief examples of how the citizen science approach of the NASA GLOBE Program supports the UN SDGs are detailed in the following.

SDG 13: Climate Action

The NASA GLOBE Program supports data collection on climate-related parameters such as temperature, precipitation, and surface temperature. NASA GLOBE engages its participants in the long term monitoring and the understanding of climate change impacts on their communities. Using these data is in support of efforts to mitigate and adapt to climate change while at the same time fostering a sense of environmental stewardship and promoting sustainable practices.

SDG 14: Life Below Water

NASA GLOBE participants can play a crucial role in monitoring marine and freshwater ecosystems by documenting the biodiversity of these systems and identifying threats such as water pollution and the degradation of habitats. The data that participants collect and share around these aquatic environments can serve as a crucial data store to inform conservation efforts and the sustainable management of marine resources, which aligns with the targets identified in the objectives of SDG 14.

SDG 15 Life on Land

We have highlighted in previous chapters the incredible sources of remotely-sensed data and we have at our disposal. Because of the scale of these data and the aims of SDG 15, the protocols that the NASA GLOBE Program offers its participants align very well. NASA GLOBE participants can monitor biodiversity across the landscape as well and be able to monitor the impacts of habitat restoration efforts. The data collected around species distribution and ecosystem health are incredibly important. Additionally, these data can help monitor changes in overall land cover distribution and the impacts of human decision making around land use on the local community. These are important tenets to the preservation of terrestrial ecosystems and in the promotion of sustainable land use practices.

SDG 3: Good Health and Well-Being

The work of NASA GLOBE is often used to focus on and address public health issues by engaging communities in disease surveillance, air and water quality monitoring, and epidemiological research around vector-borne disease. These data collected by NASA GLOBE participants are serving to empower citizens to participate in health-related data collection and analysis, these initiatives contribute to promoting well-being, preventing diseases, and informing evidence-based policymaking aligned with SDG 3. An important example of this is the work around the identification of mosquito populations and monitoring of species known to be vectors of disease such as Zika in communities around the world.

SDG 11: Sustainable Cities and Communities

Globally there is a shift of populations from rural areas to urban centers. The NASA GLOBE Program has initiatives that are centered on data collection around air quality and the growing challenges around heat related illness in these urban settings. In monitoring air quality and the impact of urban infrastructure on climate-related variables, NASA GLOBE participants are collecting important data around important public health indicators. By empowering citizens to collect data on local environmental conditions and advocate for sustainable urban development, these projects can contribute to creating inclusive, safe, resilient, and sustainable cities, in alignment with SDG 11.

Overall, the focus on citizen science that NASA GLOBE advocates serves as a powerful tool for advancing the UN SDGs by mobilizing individuals and communities to actively participate in scientific research, environmental conservation, public health initiatives, and educational outreach efforts.

Central to the NASA GLOBE Program are its overarching objectives and framework, which guide its activities and initiatives. This mission of NASA GLOBE is to promote scientific inquiry, environmental stewardship, and cross-cultural collaboration. The core components of the NASA GLOBE framework are the development of scientific protocols, data collection parameters and community engagement. The scientific protocols are developed by the lead scientist(s) that are seeking the collaboration of the NASA GLOBE community. These rigorous protocols are then developed in a manner that allows for non-scientists to implement in a way that ensures that the rigor of the protocol is maintained through high levels of efficacy. The data collection parameters allow these data to be geolocated which then allows them to serve as in-situ validation of data that is collected remotely. In implementing the scientific protocols and data procedures communities are engaged in the process that offers the important local perspective needed to help address issues.

The NASA GLOBE program has truly been a success story in the infusion of inquiry-based STEM education across the globe. NASA GLOBE has, across its three decades in existence, fostered an incredible record of meaningful environmental learning experiences from the very focused to local needs and interests to large-scale global targeted campaigns that have looked at broad scale issues like the urban heat island effect exacerbated by a changing climate. This has been accomplished through sustained professional development support of those involved and the ongoing development of resources to support collaborative research across the community. This multifaceted approach of supporting the community serves as an incredible example for those looking to foster similar partnerships around their work.

The NASA GLOBE program has focused on measuring the impact of their initiatives. Measuring the impact of the NASA GLOBE Program entails assessing its outcomes on various stakeholders, including students, educators, scientists, and policymakers. Based on this work, it is broadly recognized that the NASA GLOBE program has had a significant impact and influence on the development of environmental literacy. The NASA GLOBE program should be applauded for its focus on using a mixed-method methodology approach for measuring its impact. NASA GLOBE has used the range of measures from empirical studies, to qualitative assessments, to the use of anecdotal evidence to gauge the program's influence on the development of environmental literacy, its influence on shaping STEM education, the role it plays in scientific research, and levels of community engagement as an outcome of being involved in the implementation of the program.

NASA GLOBE and Global Competence

The NASA GLOBE Program plays a crucial role in fostering global competence among students, educators, and communities worldwide.

Global competence refers to the knowledge, skills, and attitudes necessary to understand and address global issues effectively, while also appreciating cultural diversity and collaborating across boundaries. The NASA GLOBE Program accomplishes this through some very deliberate design considerations

By bringing together participants from 127 countries across the world and region, the NASA GLOBE Program provides opportunities for cross-cultural exchange and collaboration. Through joint research projects, data sharing, and virtual exchanges, students and educators interact with peers from diverse cultural backgrounds, fostering mutual understanding and respect.

Central to the GLOBE Program is the emphasis on environmental stewardship and sustainability. By engaging in hands-on environmental monitoring and research, participants in the NASA GLOBE Program develop a deeper connection to their local environment while also recognizing their role as global citizens responsible for protecting Earth's resources for future generations.

The activities used in the NASA GLOBE Program integrate various disciplines across the STEM disciplines as well as geography, social studies, and cultural studies. This interdisciplinary approach encourages students to explore complex global issues from multiple perspectives, enhancing their critical thinking and problem-solving skills.

The NASA GLOBE Program allows students to engage in authentic scientific research and data collection, addressing real-world environmental challenges such as climate change, air quality, and water quality. By connecting classroom learning to global issues, students develop a sense of agency and empowerment, realizing that their actions can have a meaningful impact on the world around them.

With a community that encompasses 127 countries, the work of the NASA GLOBE program would be impossible without the integration of technology. The NASA GLOBE Program leverages technology to facilitate communication, collaboration, and data analysis. Participants utilize online platforms, mobile apps, and geographic information systems (GIS) to collect, analyze, and share environmental data in real-time, fostering digital literacy and global connectivity.

NASA GLOBE encourages students and educators to engage with their local communities to address environmental issues collaboratively. By involving community members, stakeholders, and policymakers in GLOBE activities, participants develop leadership skills, build partnerships, and advocate for positive change on both local and global scales.

As NASA GLOBE participants interact with peers from different countries and cultural backgrounds, they gain a deeper appreciation for cultural diversity and develop intercultural communication skills. By learning about traditional ecological knowledge and indigenous perspectives on the environment, participants broaden their understanding of environmental issues and solutions.

All of these design considerations help to empower participants to become informed, empathetic, and proactive global citizens. In this way the NASA GLOBE Program contributes to building a more sustainable and interconnected world.

In conclusion, the NASA GLOBE Program stands as a testament to the transformative potential of collaborative scientific inquiry and environmental education. By fostering a global community of learners, researchers, and stewards, the NASA GLOBE Program embodies the spirit of exploration and discovery that defines NASA's mission. As we reflect on its achievements and challenges, let us reaffirm our commitment to nurturing the next generation of Earth scientists and environmental leaders through initiatives like the NASA GLOBE Program.

17 Ecorestoration and Regenerative Agriculture

Written with Hart Hagan, Environmental Reporter, harthagan.substack.com

The synergies between STEM education and open and data science suggest exciting opportunities for engaging students and mentors with community partners via Project Challenges. Rather than simply helping with sustainability efforts, we can strive for more equitable collaboration through partnering around ecorestoration and regenerative agriculture. These important areas of innovation connect with the SDGs, emerging interests in STEM education, and open and data science. They resonate with partners, students, academic programs, and priority investments related to sustainable development, making them particularly attractive for Experiential Learning.

Ecorestoration

Ecorestoration refers to a process of assisting the recovery of an ecosystem that has been degraded, damaged, or destroyed. The goal of ecorestoration is to return the ecosystem to a more natural and functional state, ideally resembling its condition before human intervention or disturbance. Ecorestoration involves a combination of active human interventions and natural processes to rebuild or revitalize ecological communities and their functions. Ecorestoration projects can range from small-scale efforts to restore local habitats, such as wetlands or forests, to large-scale initiatives aimed at landscape-level conservation.

Regardless of the size or scope of the ecorestoration project, the work begins with a detailed assessment of the historical and current state of the ecosystem, including its biodiversity, soil composition, hydrology, and other relevant factors. Through understanding the

DOI: 10.4324/9781003489337-21

ecological conditions and processes within a particular area, valuable information can be gathered regarding the structure, function, and health of the ecosystem, which can inform decision-making in environmental management and conservation. Ecological assessments may focus on species and populations, habitat quality and availability, ecosystem services such as water filtration, carbon sequestration, and pollination, and ecological interactions and relationships within the ecosystem. Ecological assessments serve as the basis for identifying conservation priorities, designing restoration projects, and monitoring the effectiveness of management interventions. They may be conducted at various spatial and temporal scales, depending on the objectives of the assessment and the scope of the ecosystem under study.

The work of ecological assessment represents an exciting area for collaboration related to Project Challenges. Community-based partners can share on the ground data related to climate change along with the monitoring of interventions. They can also provide historical information and data in the form of stories and personal accounts of the impacts of climate change that can complement information made available through open and data science. Engagement in ecological assessment provides students with exciting opportunities to leverage technology-based tools and software to create and examine maps, utilize advancements in AI, modeling, machine learning and other innovative technologies. Through collaborative projects, students and partners can strengthen the ability of ecological assessment to lay the foundation for powerful goal setting and eventual implementation of interventions and practices.

Based on the work of ecological assessment, specific objectives and desired outcomes for restoration projects can be set, taking into account ecological, social, and economic considerations. When we look at the most challenged parts of the world, and those that are particularly vulnerable to the effects of climate change, especially for marginalized populations, expectations for ecorestoration approaches must be highly practical. In addition to preventing further damage and climate vulnerability, approaches need to address community needs related to sustenance, cooking fuel, shelter, and protection. Because extreme poverty is concentrated within the most rural communities that are also the most climate vulnerable, the needs of these populations must be taken into account during assessment and the development of related intervention projects. This need provides framing context for the importance of regenerative agriculture approaches that enhance productivity of land through farming in harmony with nature.

Regenerative Agriculture

Regenerative agriculture is an approach to farming that supports surrounding ecosystems and is itself an ecosystem that tends to regenerate soil, replenish groundwater and provide space for fish and wildlife. This stands in contrast to conventional agriculture which requires intensive use of fossil fuels, chemical fertilizers, toxic pesticides and heavy equipment, thus displacing fish and wildlife, while generating pollution that is harmful to humans.

Quoting the UN on how regenerative agriculture is good for the climate:

> Climate-smart and regenerative agriculture measures designed to put farmers at the center can improve crop yields and turn farmland and pastures into carbon sinks, reverse forest loss, optimize the use of nitrogen-based fertilizers and rethink global and local supply chains to be more sustainable, reducing waste.
>
> (Strauss, 2023)

And also

> … regenerative agriculture (RA) describes farming and grazing practices that, among other benefits, absorb CO_2 by replenishing soil organic matter and restoring biodiversity to degraded soils, thereby improving the water cycle and helping to combat climate change.
>
> (Department of Economic and Social Affairs, 2023)

Regenerative agriculture lends itself to Experiential Learning because, unlike conventional agriculture, the emphasis is not on heavy equipment or large capital investment, but rather constant observation, evaluation and monitoring. This includes soil testing, plant selection, seed saving, seed swapping, and nontoxic control of pests and predators. Participants plant crops and trees, take soil samples, monitor water resources, and observe the interactions between crops, livestock, wild animals, and plant life.

Regenerative agriculture offers many benefits to local communities that meet practical needs in addition to fostering sustainability. Contrary to commonly held beliefs, regenerative agriculture can profitably produce nutritious food, especially when measuring nutrition per acre, rather than yield per acre. Regenerative agriculture can grow crops which people can sell at a premium on local markets. By contrast, industrial agriculture typically involves commodity crops, which are sold at low margins on global markets.

The environmental benefits of Restorative Agriculture are especially compelling when compared to conventional farming practices. Conventional agriculture makes heavy use of fertilizers manufactured with fossil fuels, and therefore is responsible for significant greenhouse gas emissions, while regenerative agriculture avoids the use of chemical fertilizers. Conventional agriculture typically degrades the soil, causing the soil to emit carbon dioxide as the biology decomposes. Regenerative agriculture regenerates the soil, and stores carbon in the ground, in the form of living organisms and soil organic matter. While industrial agriculture tends to degrade the soil, causing drought, desertification, and soil erosion, regenerative agriculture improves soil health, reducing erosion and improving the ability of the soil to hold water, thus reducing the impacts of drought and reversing the trend toward desertification.

Regenerative agriculture integrates trees and forests into the landscape. Therefore, it is useful to examine the effects of trees and forests on our climate. Forests and healthy soil serve to reduce flooding. This is because trees capture rainfall, and they foster a layer of organic matter on the forest floor (including leaf litter and fallen trees). This organic layer greatly reduces runoff and flooding. Furthermore, tree roots improve soil quality, making the soil much more able to absorb rainfall, reducing runoff and flooding. Conversely, a deforested area lacks the trees and the organic matter that would otherwise absorb rainfall. When trees are removed, the soil becomes compacted, because the soil then lacks the living roots that would otherwise keep the soil spongy and absorbent, reducing flooding.

Just as deforestation causes flooding, it also causes drought, because the soil is no longer able to hold water. Soil that is able to hold water tends to prevent drought and also minimize the effects of drought. Land degradation and soil degradation cause a vicious cycle of flooding and drought. When rain falls onto degraded, compacted soil, it quickly runs off, leaving the land without water. Forests and healthy grasslands hold onto rainfall like a sponge, releasing it gently into the streams and waterways. This is the type of landscape that remains moist and healthy during periods of low rainfall. Because regenerative agriculture prevents flooding and drought, it helps support surrounding ecosystems. In addition, regenerative agriculture avoids the use of toxic pesticides, which will otherwise kill off populations of insects, which are food for a range of vertebrates, including birds and mammals. Thus, regenerative agriculture supports biodiversity which according to the United Nations is our strongest natural defense against climate change. "The Earth's land and the ocean serve as natural carbon sinks, absorbing large amounts of greenhouse gas emissions. Conserving and restoring natural spaces, and the biodiversity they contain, is essential for limiting emissions and adapting to climate impacts."

Because of the above analysis, regenerative agriculture supports SDG 15 Life on Land, "Protect, restore and promote sustainable use of terrestrial ecosystems, sustainably manage forests, combat desertification, and halt and reverse land degradation and halt biodiversity loss."

In addition, regenerative agriculture supports SDG 6 Clean Water & Sanitation. "Ensure availability and sustainable management of water and sanitation for all." Typically people live on land that has been degraded by deforestation and conventional agriculture. When you restore the plant matter and soil health, plants and soil are available to capture rainfall. This tends to recharge the groundwater and make the streams flow again with relatively clean water. Thus water becomes available in higher elevations. When streams flow in higher elevations, then people can find water closer to their homes and farms, for drinking, personal use, crops and livestock. This will tend to reduce pollution in the rivers and lakes at lower elevations. When people can wash their clothes and water their livestock in streams at higher elevations, they can avoid polluting the rivers and lakes at lower elevations, increasing everyone's access to clean water.

References

Department of Economic and Social Affairs (2023). UNFF 18 panel on private sector drivers and contributions: Regenerative agriculture for the global forest goals. www.un.org/esa/forests/wp-content/uploads/2023/04/UNFF18-CN-Private-sector-240423.pdf

Strauss, Tania and Chhabria, Pooja (2023). What is regenerative agriculture and how can it help us get to net zero food systems? 3 industry leaders explain. https://climatechampions.unfccc.int/what-is-regenerative-agriculture-and-how-can-it-help-us-get-to-net-zero-food-systems-3-industry-leaders-explain/

Epilogue

Putting Our Partners on the Map

It is hard to imagine that in an age in which most of us carry a computer in our pocket the vast majority of the Earth's surface (nearly 70%) is not mapped at all or at best is mapped in a way that does not allow us to know enough to make informed decisions. This means that millions of people around the world are approaching their daily lives without accurate maps, and the maps that the world is using are missing millions of people. The implications are profound.

The work that we have described in this book is our reaction to this global challenge. Through collaborative engagement, students are working with community partners to literally put them "on the map." They are using satellite images that are available from many different platforms and sources to identify the areas where our partners are leading ecorestoration projects and interventions. And through sharing in-situ (on the ground) geo-referenced information, our partners and their communities are informing and expanding the satellite data, contributing to maps that are inherently richer and more complete. The resulting maps that are created can then be shared back with partners to verify and expand the information included. This iterative process can support solutions to the challenges that communities face, while also creating metrics to examine the impact of this work going forward.

The act of adding our partners' data to satellite maps catalyzes new possibilities for climate action and complements the amazing images of the entire Earth that we receive every three to seven days. While some of these images are impacted by variables such as cloud cover, at the large scale where our work is focused, we can see change at an incredibly intimate level. These data are in a form that allows for a range of analysis including the option for scientists to use spectral or visual inspection of the data. This optionality is so incredibly powerful given the influence of visual data on peoples' understanding and behavior. And as we work to understand the impacts of climate change toward supporting and accelerating restoration, satellite maps and data science represent our most powerful tools. The more complete, accurate and compelling our maps are, the more they can inform and drive transformative climate action.

When we consider the potential contribution of community partners related to in-situ data collection and monitoring, a range of opportunities become evident. Georeferenced data can fill in gaps, provide local context and experiences over time, and identify emerging threats and concerns related to evolving patterns or changes. As an example, partners in coastal Kenya have shared concerning patterns associated with climate change. Because of accelerating drought conditions, animals are traveling further to find grazing land and coming into contact with people living in rural communities. They are also noting recent challenges to cultivating crops and trees, even the varieties that have been especially hardy and resilient to climate crisis.

While in-situ data is critical to understanding climate related patterns and threats, it is equally important for understanding the impacts of ecorestoration approaches. By engaging local communities and youth to monitor changes associated with regenerative agriculture or other sustainability projects, we can add data that allows for the detection of nuanced patterns and optimization for specific regions and climates. As satellite and in-situ data coalesce, and as more ecorestoration projects and approaches are implemented throughout the world, integrated maps will become increasingly powerful in informing optimal paths toward sustainability and regeneration.

But here is the most exciting part. With Project Challenges and PEARL, our students can do more than help improve climate maps. They can work with community partners to implement and monitor ecorestoration interventions while utilizing technology-supported tools to optimize success and sustainability. Through this work, we can cultivate a sense of hope and motivation as we recognize our power, and the inherent potential of youth, community partners, and higher education to mobilize our individual and collective resources.

Appendix

Featured Technology-Supported Tools and Resources

Open Geospatial Technologies

Open geospatial technologies encompass a wide array of tools and methodologies that leverage open-source software, data, and standards to analyze, visualize, and manage spatial data. These technologies play a crucial role in various fields such as urban planning, environmental monitoring, disaster management, and agriculture.

One key component of open geospatial technologies is open-source Geographic Information Systems (GIS) software, such as QGIS (www.qgis.org) and GRASS GIS (www.grass.osgeo.org) These platforms provide powerful capabilities for spatial data analysis, visualization, and mapping, without the high cost associated with proprietary software.

Open geospatial technologies also rely heavily on open data sources, including satellite imagery, aerial photographs, and geospatial datasets collected by government agencies and research institutions. These datasets are often freely available for download and use, enabling researchers, developers, and policymakers to access valuable spatial information without restrictions.

Interoperability is another important aspect of open geospatial technologies. Open standards like the Open Geospatial Consortium (OGC) standards ("Standards," n.d.) ensure that geospatial data and software can work together seamlessly, regardless of the platform or software used. This interoperability facilitates data sharing and collaboration among different stakeholders.

Web mapping technologies, such as Leaflet (www.leafletjs.com) and OpenLayers (https://openlayers.org), are widely used in open geospatial applications to create interactive maps and web-based spatial applications. These technologies enable users to visualize geospatial data online, share information with others, and develop custom mapping solutions tailored to specific needs.

Open geospatial technologies also support crowdsourced mapping initiatives, where volunteers contribute data to create and update maps in real-time. Projects like OpenStreetMap (osm.org) rely on contributions from individuals worldwide to create a free and editable map of the world, which can be used for various purposes including humanitarian aid and urban planning.

Spatial databases, such as PostgreSQL (www.postgresql.org) with PostGIS extension, are essential tools in open geospatial technologies for storing and managing large volumes of spatial data efficiently. These databases support spatial queries, indexing, and geospatial functions, making them well-suited for handling complex spatial datasets.

Machine learning and artificial intelligence are increasingly being integrated into open geospatial technologies to automate processes such as image classification, object detection, and spatial analysis. These technologies enable faster and more accurate extraction of information from geospatial data, leading to new insights and applications.

Cloud computing platforms, such as Amazon Web Services (AWS) (https://aws.amazon.com) and Google Cloud Platform (GCP) (https://cloud.google.com), provide scalable infrastructure for hosting and processing geospatial data in the cloud. This enables users to access computational resources on-demand, reducing the need for expensive hardware and software licenses.

Overall, open geospatial technologies democratize access to spatial data and tools, empowering users from diverse backgrounds to explore, analyze, and contribute to the understanding of our planet's geography and environment. By leveraging open-source software, data, and standards, these technologies foster innovation, collaboration, and transparency in the geospatial community.

Open Access Satellite Imagery

Open access satellite data refers to satellite imagery and related data that is freely available for anyone to access, download, and use without restrictions. This wealth of data is made possible by various government agencies, research institutions, and commercial entities that share their satellite observations to promote scientific research, environmental monitoring, disaster response, and economic development.

One of the primary sources of open access satellite data is government-operated satellite programs, such as NASA's Earth Observing System (EOS) (https://eospso.nasa.gov/content/nasa-earth-science-data) and the European Space Agency's (ESA) Copernicus program (www.esa.int). These programs collect a wide range of Earth observation data using satellites equipped with sensors capable of capturing imagery at different wavelengths and resolutions.

Satellite imagery provided by these programs covers various aspects of the Earth's surface, including land, oceans, atmosphere, and cryosphere. This data is invaluable for studying phenomena such as land cover change, deforestation, urbanization, sea level rise, and climate change, among others.

In addition to government-operated programs, there are also commercial satellite operators that offer open access to some of their satellite imagery. Companies like Planet (www.planet.com) and DigitalGlobe (www.maxar.com) provide high-resolution satellite imagery to the public for free or at a reduced cost through initiatives like the Amazon Web Services (AWS) Public Dataset Program.

Open access satellite data is typically distributed through online platforms and data portals, where users can search, preview, and download datasets based on their specific needs. These platforms often provide tools for data visualization, analysis, and integration with other geospatial datasets.

One of the key benefits of open access satellite data is its democratizing effect on Earth observation. By making satellite imagery freely available, researchers, educators, NGOs, and governments around the world can access the same data, regardless of their financial resources or geographic location.

Open access satellite data plays a critical role in disaster response and management. During events such as hurricanes, wildfires, and earthquakes, satellite imagery can provide timely information on the extent of damage, aid in search and rescue efforts, and support decision-making by emergency responders and relief organizations.

Furthermore, open access satellite data enables innovative applications and services in areas such as precision agriculture, urban planning, transportation, and natural resource management. By integrating satellite imagery with other data sources and technologies like GIS and machine learning, new insights and solutions can be developed to address complex challenges.

Despite the availability of open access satellite data, challenges remain in terms of data quality, processing capabilities, and access to high-resolution imagery for certain regions. Efforts are underway to address these challenges through improved data standards, collaboration among stakeholders, and the development of advanced analytical tools and algorithms.

Open access satellite data has revolutionized the field of Earth observation by providing unprecedented access to valuable information about our planet. By leveraging this data, researchers and practitioners can gain insights into environmental processes, monitor changes over time, and inform evidence-based decision-making for a sustainable future.

Reference

Standarids. (n.d.). Open Geospatial Consortium. Retrieved April 21, 2024, from www.ogc.org/standards/

Index

Note: Page numbers in *italic* refers to Figures.

www.ingramcontent.com/pod-product-compliance
Lightning Source LLC
LaVergne TN
LVHW011710070225
803225LV00003B/104

* 9 7 8 1 0 3 2 7 4 1 7 1 0 *